# The Botanical Odyssey *of* Richard S. Felger

The Southwest Center Series

JEFFREY M. BANISTER, EDITOR

EDITED BY
BENJAMIN T. WILDER, BILL BROYLES,
AND THOMAS BOWEN

# The Botanical Odyssey *of* Richard S. Felger

THE UNIVERSITY OF ARIZONA PRESS
TUCSON

The University of Arizona Press
www.uapress.arizona.edu

We respectfully acknowledge the University of Arizona is on the land and territories of Indigenous peoples. Today, Arizona is home to twenty-two federally recognized tribes, with Tucson being home to the O'odham and the Yaqui. The university strives to build sustainable relationships with sovereign Native Nations and Indigenous communities through education offerings, partnerships, and community service.



ISBN-13: 978-0-8165-5725-7 (paperback)
ISBN-13: 978-0-8165-5726-4 (ebook)

Cover design by Leigh McDonald
Cover photograph of Richard Felger in the Gran Desierto by Bill Broyles
Designed and typeset by Leigh McDonald in Granjon 10.5/14 and Brandon Grotesque (display)

Publication of this book is made possible in part by support from the Southwest Center of the University of Arizona.

Library of Congress Cataloging-in-Publication Data
Names: Wilder, Benjamin Theodore, editor. | Broyles, Bill, 1944 July 9– editor. | Bowen, Thomas, 1942– editor.
Title: The botanical odyssey of Richard S. Felger / edited by Benjamin T. Wilder, Bill Broyles, and Thomas Bowen.
Other titles: Southwest Center series.
Description: Tucson : University of Arizona Press, 2026. | Series: The Southwest Center series | Includes bibliographical references.
Identifiers: LCCN 2026005258 (print) | LCCN 2026005259 (ebook) | ISBN 9780816557257 (paperback) | ISBN 9780816557264 (ebook)
Subjects: LCSH: Felger, Richard Stephen. | Botanists—United States—Biography. | LCGFT: Biographies.
Classification: LCC QK31.F39 (print) | LCC QK31.F39 (ebook)
LC record available at https://lccn.loc.gov/2026005258
LC ebook record available at https://lccn.loc.gov/2026005259

Printed in the United States of America
♾ This paper meets the requirements of ANSI/NISO Z39.48-1992 (Permanence of Paper).

# CONTENTS

Richard Stephen Felger, 1989. Richard Felger archive.

# PREFACE

RICHARD FELGER (1934–2020) was one of the most important scientists working in the American Southwest and northwestern Mexico in recent generations. Though he was best known as a Borderlands botanist, his interests and expertise reached far beyond botany and regional geography. He was a global visionary, an uncompromising advocate for the conservation of biological diversity and the search for new resources to feed an overpopulated planet. He was a tireless worker, and a paragon of thorough and meticulous research. Many of his books and papers are acknowledged by his peers as masterpieces.

Richard was also an extraordinary human being. He traveled widely, was an astute observer, and had a profound appreciation of and empathy for people beyond our borders. He had a great many friends and colleagues who were deeply fond of him, though nearly all found him to be more than a bit eccentric. His friends and colleagues also knew him as a gifted writer, not only for the simple clarity of his scientific writing, but for his consummate storytelling.

This book is both by Richard Felger and about him. In part 1, Richard adopts his autobiographical storytelling persona, telling about some of his wildest as well as his most sublime field and travel experiences, with a sly humor and masterful prose. In part 2, Richard's friends and colleagues relate with love and exasperation what fieldwork and co-authorship was like with this supremely dedicated botanist who never had enough time to look for plants or enough presses to preserve them. Part 3 explores his family history and traces how his unconventional personal and professional life unfolded. The book concludes with a bibliography of Richard's books and his most important papers and articles. Together, the four parts present a portrait of one of the great scientific personalities of our time.

# ACKNOWLEDGMENTS

DESPITE HIS BEST efforts, Richard was not able to complete this book during his lifetime. After his death, a draft manuscript of Richard's stories from the field was transferred to Ben as part of Richard's larger archive. Over the years, Richard had shared a few of these stories with the three of us, and once we—Ben, Bill, and Tom—became aware of the full scope of these accounts, we realized it was a work that needed to be completed. We envisioned a book that would consist of not only Richard's own narratives, but accounts of life with Richard in the field, in the herbarium, and on the computer as told by him and those of us who knew him best.

The outpouring of enthusiasm for the project was truly gratifying. It is a fitting tribute to Richard that nearly everyone we invited to contribute a story accepted our invitation, and their highly varied and individual accounts help round out the picture of Richard as only his friends and colleagues can. We thank our eighteen contributors for their excellent chapters. It has been a pleasure working with such a wonderful group of desert rats to bring this book to fruition.

Many individuals have contributed to the publication of this book in a variety of other important ways, and we wish to take this opportunity to thank Adrienne Booth, Mahina Drees, George Ferguson, Nick Georgiou, Gil Gillenwater, Diana Hadley, Jonathan and Roseann Hanson, Gene Joseph, Angelica Leyva and Marisa Quiroz at the International Community Foundation,

Melanie Mallon, Shelley McMahon, Gary Nabhan, Erin Riordan, Sue Rutman, Paul Tilley, Robert Wegener of Statistical Research Inc., Sula Vanderplank, Robert Villa, and Janos Wilder.

Thank you to Jeffrey Banister of the University of Arizona's Southwest Center, and Jennifer Jenkins, for their belief in this book and for helping us find its rightful home at the University of Arizona Press, where Kathryn Conrad and Allyson Carter and their team gracefully and generously made this book become a reality.

We also thank three anonymous reviewers for their helpful and constructive comments, which both encouraged us and strengthened the manuscript.

We will be forever grateful to Silke Schneider, Richard's wife, whose generosity and continuous support illuminated lesser-known aspects of his life.

The opportunity to meet or become even closer friends with these individuals has been one of many gifts from this project. Thank you all.

# The Botanical Odyssey *of* Richard S. Felger

Arizona, Sonora, and the Baja California peninsula.

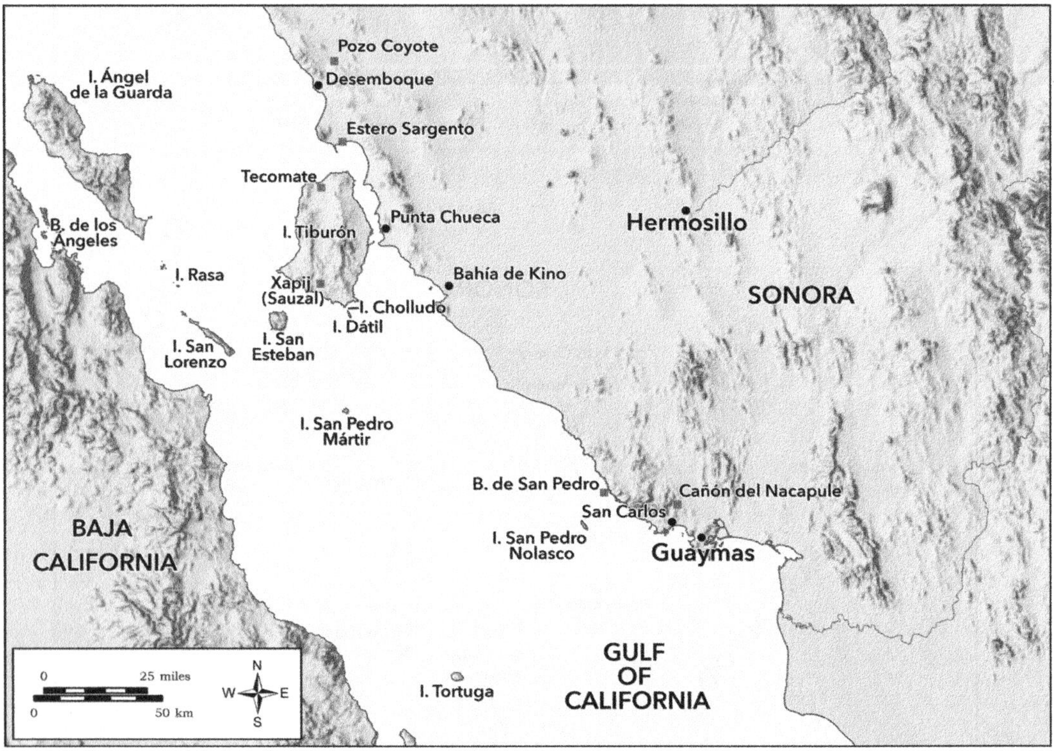


The Midriff Islands region of the Gulf of California, Mexico.

Southwestern New Mexico.

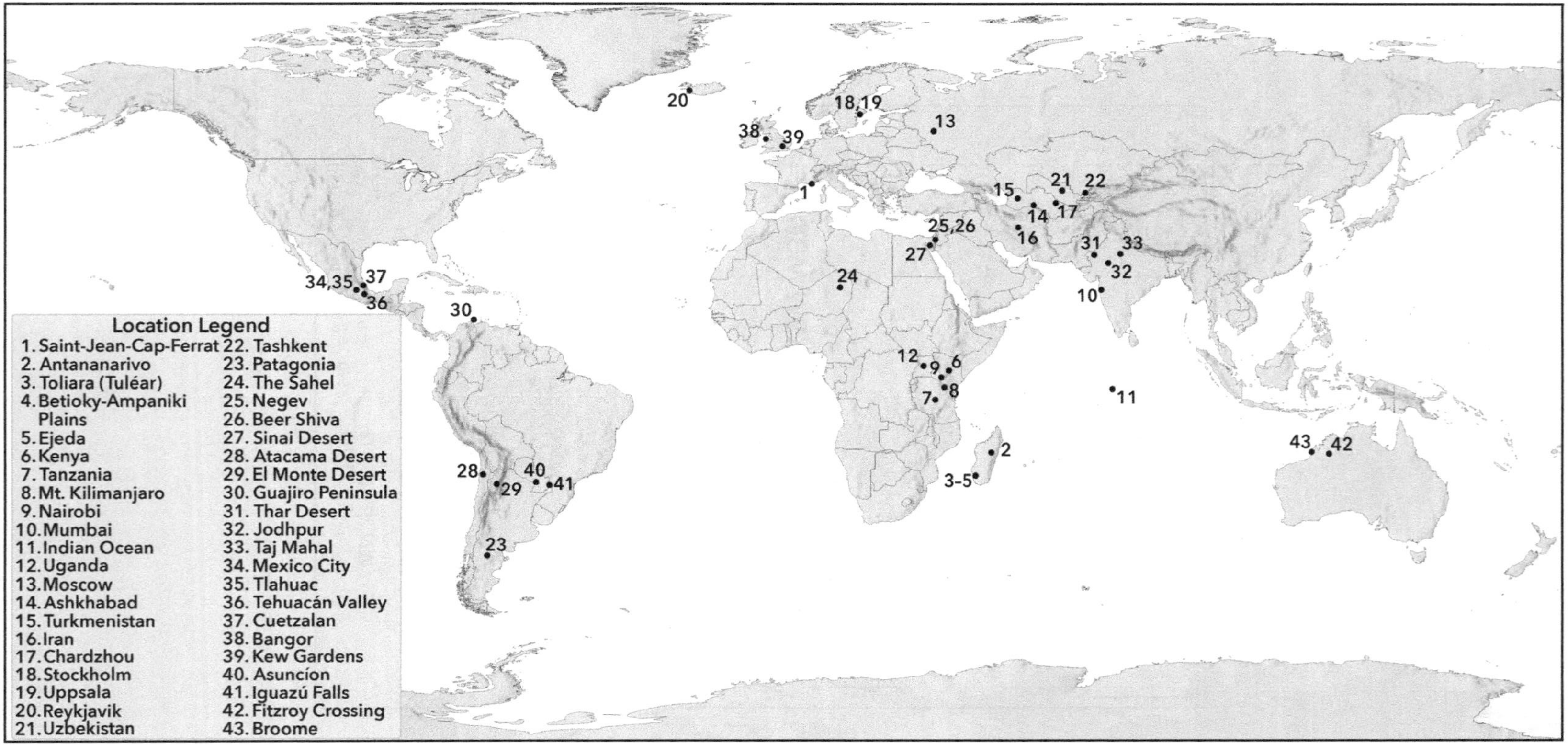
Location Legend
1. Saint-Jean-Cap-Ferrat
2. Antananarivo
3. Toliara (Tuléar)
4. Betioky-Ampaniki Plains
5. Ejeda
6. Kenya
7. Tanzania
8. Mt. Kilimanjaro
9. Nairobi
10. Mumbai
11. Indian Ocean
12. Uganda
13. Moscow
14. Ashkhabad
15. Turkmenistan
16. Iran
17. Chardzhou
18. Stockholm
19. Uppsala
20. Reykjavik
21. Uzbekistan
22. Tashkent
23. Patagonia
24. The Sahel
25. Negev
26. Beer Shiva
27. Sinai Desert
28. Atacama Desert
29. El Monte Desert
30. Guajiro Peninsula
31. Thar Desert
32. Jodhpur
33. Taj Mahal
34. Mexico City
35. Tlahuac
36. Tehuacán Valley
37. Cuetzalan
38. Bangor
39. Kew Gardens
40. Asuncíon
41. Iguazú Falls
42. Fitzroy Crossing
43. Broome

The world.

PART I

# NOTES FROM THE FIELD

Richard Stephen Felger (1934–2020)

# 1

# A SAPLING GROWS UP

I THOUGHT ABOUT lots of things walking to Third Street grammar school and then bicycling to John Burroughs Junior High on sycamore-shaded sidewalks in mid-twentieth-century early smog days. Our street trees were European sycamores, the London plane, even though our street was named Las Palmas. I would have preferred California sycamores, even though the trunk grows too crooked for a normal street tree. Native trees for native microcosms reaching native birds.

I imagined all the people gone, as they will be someday. Trees and vines growing through the ruins. Most non-natives will vanish from lack of artificial water. I figured which ones would reproduce. Natives would return in stages of succession, even though I did not know there was an ecological theory of succession. There were still vacant lots sporting native plants, although the majority of the Los Angeles flora consisted of weedy Eurasian invasives. Sometimes I thought about bringing extirpated mountain lions and condors and resurrecting Ice Age La Brea Tar Pits creatures—ground sloths, dire wolves, saber-toothed tigers, and mammoths. What would it be like to see muscular saber-toothed tigers in action? Or might they use their sabers to rake rotting flesh off mammoths mired in the tar? What's wrong with a heroic animal being a scavenger? America's national symbol is a carrion eater and so is Mexico's.

The rundown house that looked like a miniature castle, where a young movie star once lived, had Italian cypress trees. Tall dark columns, green on the outside and itchy scratchy dead twigs inside full of spiders and mites. Those trees would live on in this Mediterranean climate but would not spread. I never saw seedlings and later learned they are a sterile horticultural selection, cloned by cuttings. Olive trees, also cloned Mediterranean selections, would keep going, some for centuries or a thousand years, but I never saw seedlings. Each tree produces thousands of one-seeded olives every year, so maybe a few would germinate and survive and admit olives to a future flora.

Our neighbor, a world-renowned allergist, had a whole grove of olive trees providing the neighborhood with hay fever. Olive pollen is a top-listed allergen. A favorite trick was to give fresh juicy olives to a new kid from back East. Unprocessed olives will pucker your mouth for a long time.

Sidewalk edges and street cracks would fill first with weedy non-natives like Bermuda grass, but to make things more interesting, I imagined natives like coastal prickly pear (*Opuntia littoralis*) and Los Angeles manzanita (*Arctostaphylos glauca*). They would thrive in cracks opening from earthquakes. Being a California kid, I thought earthquakes were normal. Native and naturalized species could move in from vacant lots and big open spaces like the La Brea Tar Pits, about a mile away, still vaguely natural. The tar pits were our hunting grounds. Most were not fenced, incredible as it might seem in our litigious society. After spring rains, we found California tree frogs breeding in water topping off tar pools. Two-striped garter snakes lurked for fat little frogs. Some years later the Los Angeles County Museum of Art and the Page Museum of tar pit fossils and their parking lots smothered our hunting grounds.

Not far beyond school was a stream with native sycamores, California live oaks, and big California newts (*Taricha torosa*). Unlike most salamanders, they have dry rough skin. Big eyes, fixed smile, and rising up inquisitively all make them look happy. Fun to give one an earthworm and watch it stuff down a wet wiggling worm with its yellow-bottomed splayed-out hands. I was in junior high when the stream was buried in a concrete coffin—my first experience of a world raging against nature, against my world. I was waking up to deceits like the engineering marvel of robbing Owens Valley water for Los Angeles. Someday the land will recover after people leave. Daydreams of paradise regained, although I did not have a paradise lost until the death of my California newts and their shady stream.

Teachers did not care much about animals and plants, marine mollusks and other sea creatures, let alone cycads, cactus, and orchids. I made it through school days mentally evolving landscapes and morphing creatures and plants. One favorite was a basic iguanid like a fence lizard, which I envisaged evolving into something more specialized like a chuckwalla, horned lizard, or fringe-toed sand swimmer. I was more closely tied to the Pleistocene than the Mesozoic, so dinosaurs didn't hold much interest. Turtles, fish, cactus, cycads, palms, lizards, mollusks and other sea life inhabited my secret territory. I wondered how other kids got through. Good that we had windows so I could see trees for new ideas. I would maneuver a seat in the back of the room so I could read an interesting book hidden inside a dumbed-down schoolbook.

Besides the tide pools a mile or so from the beach house we rented in the summer, favorite places the rest of the year included Joshua Tree forest in the nearby Mojave Desert and the Natural History Museum downtown. Old Dr. Hill was curator of marine invertebrates. He befriended any kid interested in marine mollusks (what everyone else called seashells) and helped establish the local shell club and interest in marine invertebrates. I helped label and file specimens, each accession in a museum-special small open-top cardboard box snuggled into wide wooden pullout drawers. It was exhilarating to see known species, like old friends, and examine new ones, learning names and seeing related species in adjoining boxes—easy to visualize evolutionary connections. The specimen labels were intriguing, tying specimens to habitats and localities such as intertidal, sandy beach, rocky shore, or mysterious deep-water fathoms, Southern California, Baja California, the Gulf of California, and South Sea Islands from shells donated by WWII veterans. I was appalled that some kids would steal shells. Old Dr. Hill had to know. And then I got to take a Saturday natural history class at the museum. For the first time I was learning something worthwhile in a classroom. Dr. Gretchen Sibley was head of the Saturday natural history classes. I can see her even now. Did she know how much I loved those classes? Her specialty was the structure and identification of mammalian hair. The subject was not of much interest to me, but for the first time I realized it was okay to follow your interest, even if the rest of the world, including your own family, did not share the intrigue.

I read every book on natural history and ecology that I could get my hands on. Among the most treasured were Raymond Ditmar's reptile books and the three-volume small-print thin-paper *Cyclopedia of Horticulture* by Liberty Hyde Bailey that I pestered my father to buy for me. The bookstore salesman ignored me and told Dad it was not a book for children. For the first time I confronted adult authority in anger. I was ten years old and got my cyclopedia.

Later I got Archie Carr's sea turtle books and Steinbeck and Ricketts's 1941 *Sea of Cortez*. I had no interest in the Steinbeck part, only Ricketts's vivid descriptions and photos and marine life. That book set me on a course of discovery to the Gulf of California. I still have those books and continue to use Bailey's *Cyclopedia of Horticulture*, dating from 1900 to 1947. Look at the four thousand intricate line drawings done by LHB himself. He was still collecting palms in the Amazon in his late nineties. Later in life I met Archie Carr, and he did not disappoint me. He called me Chico. I realized it was a good way to avoid remembering someone's name—although we met more than once in Central America for sea turtle conservation, he still called me Chico.

A major turning point was my high school biology teacher, Nancy Thomas Neeley. She encouraged me. Well, my parents tried, in their way, but since they didn't understand what motivated me it did not mean much, and besides, kids don't give parents much credit until later in life. Maybe I got interested in natural history because my father let me roam the seashore and mountain wilds as I pleased and generally ignored me, and my mother was unable to do things that held my interest. But as E. O. Wilson points out in *Biophilia*, there is more to it than that. I just plain took to animals and plants and wild places. I remember when a friend of the family, Albert Harris—he was a soldier then—gave me a birthday present box of cactus and succulents, exotic shapes that mirrored shapes I knew from the sea. I was eight years old. I planted them in a specially prepared bed. I brought in fresh soil since the sand in the front yard of the beach house was too salty. Those plants thrived, even after I transplanted them to a more permanent place at our home in town. That was the turning point into botany. Until then I was interested only in marine biology.

Back to Nancy Thomas. At high school lunchtime I hung out in Nancy's classroom with natural history friends. It was a wooden bungalow that had been moved onto the school grounds. Sanctuary from the sports-inflicted outer world. Scientific names only, no common names here. Nancy brought us scientific journal articles, some from her boyfriend Peter Neeley, who was working on his PhD at UCLA. After class Nancy often took us in her car to the Palos Verdes tide pools, and every new creature encountered and identified added to real education. Through Peter I came into contact with UCLA professor Raymond Cowles. I went to seminars, tagged along on field trips, and listened to graduate students talking about their ecological research.

Professor Cowles told me about *Eugenia* trees in his native South Africa that took the place of California oaks. We had a tall *Eugenia* hedge around the backyard to blot out neighbors. After a bit of reading, I realized our hedge could grow into magnificent trees if they could escape pruning shears. I never liked plants pruned into bondage. From Dr. Cowles and his students, I learned about thermoregulation, how the delicate little *Xantusia* night lizards that lived under decaying fallen Joshua tree branches thrived in the desert even though they would perish in minutes in the summer sun only inches away from their microhabitat. Dr. Cowles told us about South African racial policies that he detested and population bomb science years before Paul Ehrlich's masterpiece.

One Christmas vacation, Nancy and Peter, now married, invited me to go along on a trip to Álamos in southern Sonora. It was the subtropics beyond the desert. Not that the desert, first the Mojave and later the Sonoran Desert, did not have a strong appeal. But the edge of the tropics, the edge of everything that makes dreams real, opened my world: bromeliads, cycads, palms, orchids, towering cacti, long-nosed green vine snakes, boa constrictors, cichlid fish, and tropical trees. Such splendors made Southern California sclerophyllous scrub seem dull and deprived. I know there is special purpose for the fire-climax California coastal flora, but it never pulled on my soul like the tropics.

Richard (*far right*) exploring the Sierra de Álamos and orchid-clad boulders after arriving at the University of Arizona, late 1952 or 1953. Richard Felger archive.

The tourist draw for Álamos was a colonial town in Old Mexico, cobblestone streets and falling arches. Hardscrabble poverty on back alleys smeared out in quaintness.

There were UCLA and natural history people traveling in three vehicles. Occupants of one of the cars were Audubon birders. Their car looked like an upside-down bathtub and was about as fast as a bathtub. We stopped often to look for birds, trees, and other special events, and for meals or to camp. You could camp just about anywhere; personal safety was not an issue.

The birders were Catholics and did not believe in birth control. A rude awakening, that biologists could be anti-population control. I argued that any habitat—and after all, the whole world is just one big habitat—can support only so much of any one species, or of all life itself, and no more. They were not buying it. I would say we will have population control or death control. They said we need to make more room at the table. Birders should know better. I was fifteen and sure of myself.

We went round and round, human population exploding, the environment suffering. More and more people needing more and more resources, wiping out more and more species and each other. I argued that escalating population correlates with warfare and violence, increased violence to the natural world and fellow humans. I pointed out that we didn't have atomic bombs until population reached two billion. They lamented the destruction and loss of natural habitats and species, what we now call biodiversity, and claimed to be pacifists as well as birders. But in the next breath they would say we have to make more room at the table. Did you think that one up or did you get it from the biologically challenged Pope? What if you can't get enough wood to make a bigger table? Well, just make more room at the table and share. Yeah, but how many times can you divide the pie and how long can you still find trees to make a bigger table? Round and round. Everyone else got tired of my arguing, although they agreed with me.

Because of my introduction to the subtropics in Sonora, I decided on the University of Arizona, since it was the closest I could get to the tropics. And I subsequently spent much of my life in Tucson, my home base in the Sonoran Desert. Undergraduate and graduate days, and as often as possible, fieldwork in Arizona and Sonora, and on islands in the Gulf of California. My major professor was a strong influence. The first years were good, but he became increasingly difficult and controlling, published my work in his name and other abuses. I will refer to him as the evil professor.

When I was a kid I vowed to live where there is no frost. Most people won't appreciate the significance, but if you are growing interesting plants, orchids, cycads, bromeliads, palms, and the like, then you can understand the importance of frost-free living. One summer when I was in graduate school and about to quit because of the evil major professor, my sweet wife and I drove from Arizona across the South to check out southern Florida and Key West, one of the few frost-free places in the United States. Real estate prices too high, crummy soil, mosquitoes, sticky hot, and hurricanes. We were appalled by separate drinking fountains and separate restrooms. The place too flat, too green, and too wet for a desert rat, even for a desert rat dreaming of frost-free living. South-facing slopes at La Jolla, Coronado Island, and a few other places in Southern California were also candidates, but real estate prices were off the chart. A few years later, after regaining my freedom and finishing graduate school, I decided on Perth in Western Australia.

The Vietnam War was raging; I was in a country engaged in genocide and ecocide in a trumped-up war. I had my Australian immigration papers and was ready to leave when a job offer came through at the University of Colorado in Boulder. It was an interesting place with the Rockies so near. The war escalated. We protested to no avail. A professor in my department spit on me as we stood in silent protest. One faculty meeting descended into a fight; peace versus war, teaching versus research. I walked out to join a peace march. The FBI was spying on my friends, and perhaps on me too. Australia seemed like a good idea.

But then I was offered a sweet job at the Natural History Museum in Los Angeles, and with nothing else on the plate I took it. Strange to return to the place of childhood wonderment and enlightenment. Good people and good pay, but for me the place was dead and the smog tough on my system. And I am allergic to freeway traffic. The war continued to escalate. In one march we were 200,000, even though the police menace was real, since American soldiers had just shot American students at Kent State. The spelling changed to Amerika.

I was senior curator at the Natural History Museum in Los Angeles in charge of botany. I got thickly involved in publicizing population and environment. Public awareness was surfacing, and the word *ecology* crept out of scientific journals. *Ecology*, the study of the environment, became a noun for environment. Being any shade of green was still not politically viable. But that was changing, and I suppose I was part of the scene. Resistance, however, came from unexpected places. The director of the museum called me to a hands-folded meeting in his ornate office and reminded me that my salary came from the public sector. Good, all the more reason to alert the public about population and the environment. What else should a researcher at a natural history museum be doing?

Once I was invited to a lunch meeting of wealthy supporters of population issues. Because of my public statements and media coverage, I was the main attraction. An overfed gentleman ranted about the dire outcome of an overpopulated future. He waved a frozen soy burger and threatened that we would all have to be vegetarians. I felt like saying, "Do I get fresh mushrooms on it?"

Old General Draper—William Henry Draper Jr.—was a founder of the elitist Population Crisis Committee. He made me a member although I didn't know what it entailed. He had ideas of what I should be researching. I should make an inventory of all the environmental problems in the world, all the pollutants. He made the connection between population and pollution and the environment but was not interested in a connection between population and warfare and weapons of mass destruction. I let my membership fade when I learned of his racist eugenics ideas.

Joe Pyne, a smashmouth specializing in making fools of publicity-seeking guests, had a national TV show. He wanted me on his show after the *Los Angeles Times* headlined an interview with me, quoting out of context that if we didn't do something about population, civilization would collapse in fifty years. Joe kept pestering me to be on his show. I mentioned it to Paul Ehrlich, and he had the hots to be on Joe's show. So I told Joe to get Paul instead of me. I finally relented if both of us would be on since I was concerned about Joe's behavior. I needn't have been concerned. Paul hogged the whole time; he might just as well have put his hand in front of my face. I didn't care. Paul was as big an egomaniac as Joe, the difference being Paul was brilliant and had a lot to say. Joe Pyne is forgotten, and Paul Ehrlich is not.

Paul's book *The Population Bomb* exploded across the planet. The Left vilified him as an elitist pig, the Right vilified him as antibusiness, and academics vilified him for nonscientific writing. Later detractors smirked that many of his predictions had not happened. Or have they? Nuclear winter became another of Paul's insights that was ridiculed and luckily has not happened, although Three Mile Island, Chernobyl, Fukushima, and thousands of nukes-in-waiting make Ehrlich's books ever more momentous. *The Population Bomb* is the most important book of the twentieth century.

The Los Angeles County Museum of Natural History was a wondrous place. The building was about half a mile long. Monday mornings, when the public wasn't there and the halls were dark except the exhibits, it was more unreal than any movie set. One of the guards ran a drug and prostitution business out of the back door and asked if there was anything I wanted, no charge because I was one of the few curators who talked to him. He repeated the part about "anything."

One day Joe the photographer said he wanted to show me something, and we went down into the lower bowels of the building, down and down, through locked doors until we were in the middle of the wind tunnel. The museum was built with a natural cool air intake and exhale system in the days before smog and air-conditioning. The wind tunnel was tall enough to walk through but not very wide and ran the length of the building. A few bare light bulbs dangled on thin cords. Brick walls painted white and a little wooden table of strange colors stood in the midpoint of the tunnel. There was a plate of candles of dripped colored wax. Joe and his friends kept their stash behind a loose brick next to the table. I am sort of claustrophobic and was relieved to get back to my office with big windows.

Curators' offices and scientific collections were on the third floor, off limits to the public. I had a wonderful office next to the herbarium, which I was in charge of. But the herbarium was not big enough to be of significance to the scientific botanical community. I drove an hour each way on the freeway in order to live by the ocean where there was less smog and the sound of waves. One day I stayed late. It was before remote alarms and beepers and such. Guards would not work after hours except with big dogs. If you stayed late you had to call down to the back door on the hour before leaving your office because of the dogs. Walking to my car, two guys came up fast and asked if I had a light. I saw their knives and ran for my car, got in just in time, and drove off. Next day I left a letter of resignation, drove out of the parking lot for the last time, as the Grand March from *Aida* came over the radio. I went back to Arizona.

# 2

# THE GUY WHO CUTS THE TOPS OFF PLANTS

WHEN MY ALLIGATORS got over five feet long, they went to the zoo and I went to college.[1] I had been to the tropical deciduous forest around Álamos with a group organized by my high school biology teacher. Palms, cycads, orchids, burseras, tree snakes, and boa constrictors. I knew where the good stuff was and the University of Arizona in Tucson was as close as I could get. At first the desert held little interest, but then one day I went to Bahía de Kino and out to Isla San Esteban with a friend from the University of Sonora in Hermosillo. The newly paved road was blocked off, so Hermosillo to Kino on the parallel dirt road was all day in clouds of powder dust enveloping my old Ford and seeping into everything. My camera didn't work after that. The fishing village at Kino had eight bars but no school. My friend wrote an indignant article for the Hermosillo newspaper.

We went to Isla San Esteban with Porfirio "Pilo" García, a tough fisherman also called "Prieto" because he was so dark. We went in his ragged little *panga* to the south shore of Isla Tiburón, and then across open, rough water to San

1. Originally published as "The Seris and the Guy Who Cuts the Tops Off Plants," in *Journal of the Southwest* 42, no. 3 (2000): 521–43. Slightly revised and reprinted with permission.

Esteban. I became fascinated with the plants and animals in desert ecosystems in primordial condition. I went back whenever possible, pressing samples of every species of plant in sight, returning to Tucson to identify them in the herbarium. One by one I identified each new plant species. Sometimes it was easy, for example, if I already knew related species in the Arizona or California desert flora. We did not have many of the excellent identification manuals now available and at times I spent a day or more just figuring out the identity of a single species.

Pilo and his brother Emilio let me go along if I helped pay for gas. One place where we camped on the south shore of Tiburón was the arroyo leading a few kilometers inland to the Sauzal waterhole. When we started off for the waterhole, Pilo took off his shoes and hid them under a bush behind the beach because he didn't want to wear out his shoes. We took water bottles to fill and he said this was where the Seri Indians got their water. The trail was littered with pottery sherds, and I noticed they were amazingly thin, averaging just 3 mm. Later I learned that Seri water-carrying vessels were among the largest and thinnest in the world.

The Bahía de Kino fishermen told stories about the Seris, and I read and heard other things, much of which I found out was false. On San Esteban, the fishermen pointed out the steep slope high above a sea cliff where long ago Seri men played a game of chicken, careening down the scree in an upturned sea turtle shell. Later I learned that Seri oral history included the same story. If a guy waited too late to bail out, the first one to yell "Your wife is mine" could take her as a second spouse. Of course this story is about the long-extinct San Esteban people whom the other Seris regarded with a bit of prejudice.

A few years later I met Jean and Ike Russell in Tucson and started going to Sonora in Ike's airplane. Wow! Just a few hours to get to Tiburón. Ike would fly low enough for me to identify plants. On one trip we landed at Palo Fierro on the east side of the island and hiked up to the mountains. I got a lot of new plant records for the island. I carried a small pair of pruning shears or clippers in my pocket and put the samples I collected in plastic bags to keep them fresh until I got back to camp. At the end of the day I would work on my notes and press the specimens—often a rather lengthy affair. This time when we got back to the plane, Ike said we should go to the Seri Indian village of Desemboque. He seemed surprised that I hadn't been there and didn't know Ed and Becky Moser.

Richard pressing plants alongside Ike Russell's airplane at Tecomate, Isla Tiburón, March 1963. Photo by Robert Thomas. Richard Felger archive.

We flew over to Desemboque in about ten minutes. Becky claims that the first thing I said when I walked into their house was "Where can I press my plants?" I suppose I was rather single minded in those days. Soon after, on another trip to Desemboque, Becky showed me some plants and asked if I knew what they were. Of course I did. She had Seri names and some uses for about fifty species of plants. I thought it would be interesting if I added scientific names and other pertinent botanical information, and we could write it up for publication in about three weeks. Twenty-five years later we finished our book on the ethnobotany of the Seris, although we published various papers along the way.

I want to pay tribute to Ed Moser, Becky Moser, Ike Russell, Jean Russell, and so many other friends who helped change my life and make it possible to record a slice of history, and of course especially to the Comcaac, or Seris, as they are known to outsiders, who likewise enriched my life and career.

I am sharing some of my experiences working with Seris, work that culminated with our book on the ethnobotany of the Seris, and some of the people

Ike Russell with his Aronca Sedan, low on gas, at Rancho Nuevo, Sonora, March 1967. Ike often relied on Mexican ranchers he knew for a few liters of Pemex to get him home. Photo by Thomas Bowen from *Backcountry Pilot*, edited by Thomas Bowen © 2002 The Arizona Board of Regents. Reprinted with permission of the University of Arizona Press.

Jean Russell next to Ike's Aronca, having just landed at Desemboque, Sonora, November 1967. Richard Felger archive.

*Left to right*: Cathy, Becky, and Ed Moser at home in Desemboque, January 1976. Photo by Thomas Bowen from *Backcountry Pilot*, edited by Thomas Bowen © 2002 The Arizona Board of Regents. Reprinted with permission of the University of Arizona Press.

and events that influenced and made that work possible. I am using the first person here since I am telling you a bit of what I did, but please keep in mind this work centered on the years of work with Seri elders, was a fully collaborative effort with Mary Beck "Becky" Moser, and closely involved her husband, Ed Moser. Their daughter Cathy contributed numerous illustrations. Certain specific studies also involved other researchers. For example, the marine algae (seaweeds) were worked on by James Norris of the Smithsonian Institution.

I started asking Seri elders about plants, often together with Becky. I asked where certain species can be found and then more and more detailed questions about plants and local natural history. It did not take long to realize that there was a vast unrecorded knowledge about to vanish. And that the elders knew a lot about plants that no botanists knew.

The kids were interested in pickup trucks and pop culture like teenagers anywhere, and there was a truncation of knowledge. Most of the older people had spent their earlier years, in some cases the first half of their lives, living off the desert and the sea. They seldom went to a store. They had almost no way to buy things. They could sell a few baskets but that was about it. Then around World War II, a road, buyers for fish products, and then fresh fish and sea turtles. They were changing from a seminomadic hunting and gathering way of life to the modern world and settling in the village of El Desemboque del Río San Ignacio and later also at Punta Chueca. Some of the Seris said it was coffee, others said it was sugar, and others said it was outboard motors and gas and oil. Just like anywhere else. No reason they should not want what we want.

So here were these fun-loving people I came to know, and it was fun to fly to Desemboque and Tiburón with Ike Russell, or drive from Tucson to the Sonora coast. It was fun to work with the Mosers, it was fun to go places with the Seris, it was fun to go to the islands with Kino fishermen and Seri fishermen. I made countless field trips to the Seri region, including a total of more than 160 days on Isla Tiburón.

I was still a student when I embarked on collaboration with Becky Moser on the Seri ethnobotany, and then I tried to keep at it through various distractions, including a few academic appointments, until I moved back to Tucson. At that time I supported my various research projects on grants and whatever else came along. Ed Janss saved the day. I wrote a grant proposal to the Janss

Foundation in Los Angeles after visiting him. Nothing happened, and I forgot about it. All other sources of funding dried up and I just about decided that working on Seri ethnobotany was economically impossible.

One day I stopped by to see Tom Bahti at his Indian arts shop, and he had a fantastic Apache burden basket, and I bought it. I brought it home and Mahina said, "It's beautiful. How much did it cost?" I told her and she said, "Now we don't have to worry about money because there's none left." As I was opening the usual rotting pile of bills, there was a letter from the Janss Foundation with a magnificent check "for Seri ethnobotany." Later, two generous grants from the National Science Foundation, a salaried position at the Arizona-Sonora Desert Museum, and support from the Desert Museum's Roy Chapman Andrews Fund, Jane Ivancovich, and many others helped make it happen.

The noted ethnobotanist Richard Ford of the Museum of Anthropology of the University of Michigan encouraged me, as did Bernard "Bunny" Fontana of the University of Arizona, and Julian Hayden, Ike Russell, and Jean Russell of Tucson, and many others. You need encouragement, especially when you are young and not sure of yourself. Beatriz "Tita" Braniff of the Instituto Nacional de Antropología e Historia in Hermosillo, and other colleagues in Mexico likewise provided help and encouragement. And of course the Seri elders were the primary encouragement.

It is important to keep in mind what it was like on the Seri coast only a few decades ago. There was no electricity, no running water, and the outside world was far away. (Later in the story there was running water of sorts in Desemboque. But when Ed Moser was not around, someone had to remember to put diesel in the pump.) The road to Desemboque was rough and slow. But you could stop and camp anywhere and be safe.

The Mosers' shortwave radio brought news, usually from Los Angeles. In their little Shangri-la house we heard of traffic delays on Los Angeles freeways and commercials for television sets and dishwashers. From Tucson I communicated with Ed and Becky by sending letters in care of a friend of theirs in Hermosillo who received their mail, and every few weeks they made the eight-hour drive from Desemboque to Hermosillo. Try as they would to explain that they were in Desemboque to work on Seri linguistics, to make it a written language, teach reading and writing in Seri and Spanish, and translate the New Testament into the Seris' language, outsiders believed the real reason was that they were working a gold mine or getting pearls.

Initially I thought that anthropologists working among the Seris must have recorded ethnobotanical information. But I soon learned that they generally skipped over knowledge of animals and plants. Apart from the serious scholars and authors, it seemed that misinformation about the Seris was rife. I often heard derogatory stories and even lurid tales of superhuman sexual practices. In fact I learned that they were rather prudish. Bride prices were steep and a girl's reputation was highly guarded.

The poverty and related problems of the Seris came to the attention of various people who tried to help them. One-on-one mutual help often made a difference. Ike Russell constantly bought things people needed and flew them to Desemboque, especially medicines. Ike's friend Dr. Bob Thomas provided medical help, and long after the good doctor was dead, mail-order discount medications bought on his name continued to go to the Seris in Ike's plane. Medical help was also available from the Mexican government, but the people had to travel to clinics in Hermosillo or the towns between Kino and Hermosillo. Ed Moser often drove people to one of the clinics or hospitals, and on occasion I did too.

One of the early efforts to help the Seris, before I arrived on the scene but still being talked about, was an American Friends Service Committee project led by Norman Kreckler. This project involved a self-help scheme for the Seris to raise chickens, but they regarded chickens as unclean animals, and that part of the project went nowhere. Fried chicken in Hermosillo was okay; they just did not think much of chickens at home, scratching about in dirt and trash. The Friends project also involved a school at El Desemboque. The teacher, Leo Sandoval, was a kind and intelligent man who I am sure contributed to the people in a positive manner. Ed and Becky arrived in Bahía de Kino in 1951.

The Seris gave creative names to the outsiders that they came to know. Even among themselves there were nicknames only used behind one's back. One Seri woman was known as Hips-a-Meter-Wide. In the early 1970s an anthropologist came to Desemboque and claimed there was a genetic disorder among the Seris that would cause imminent and widespread blindness. They named him You're Lying to Me. I was named Ctam Hehe Iyat Ctaamtim, "The Guy Who Cuts the Tops Off Plants," Ed Moser was Moosni Ilít, "Turtle

Head," and Becky was "Singing Woman" or "Blue-Skirted Woman" (she often wore blue denim skirts) or just plain Rebeca. The Seris called Ike Russell Mericaana Cacösxaj, "Tall American." There was a gringo who brought candy to the children, and as they grabbed for the goodies, he would say, "Uno, no mas." He was called Unonomas. A semi-reformed drug-addict thief was called Dead Flesh.

The Sonoran islands and most of the Seri coast were unscathed pristine desert, as if it were still the nineteenth century. The warm sea teemed with sea turtles, although it pained me to witness their populations plummeting. We could camp on uninhabited beaches and as far as you could see there was no sign of anyone else unless you looked down at the pottery sherds.

I remember early one morning in Desemboque, looking out at Seri huts along the beach, upturned wooden fishing boats, the crumbling old icehouse, and thinking how fast all this is changing. I realized I had a unique opportunity to record vanishing knowledge. I also knew that some changes would be okay and that people are not museum specimens, and tried not to romanticize too much about "the old ways." One time in 1973, Carl Hodges and a real estate developer friend of his flew with Ike and me to Desemboque. We were standing in front of town at low tide, the shallow reef exposed at sunset reflecting on the rippled wet sand, the air pungent with the smell of exposed sea life, and squawking seabirds. As I was admiring the glory of it all, the developer said, "All this land and so few people." I knew what he had in mind, and a chill crept up my spine.

I thought a lot about how the Seris could hold on to their land and not become second-class citizens in their ancestral homeland. Fishing and the sale of arts and crafts, such as ironwood sculpture and baskets, allowed them to earn income without working for someone else, which they would not do, and remain on their land. But there were political struggles and sometimes I had a difficult time avoiding being drawn into the fray. I realized that if I wanted to complete my work, work I felt nobody else would or could do, I had better stay out of politics. It was not my country and there was not much I could accomplish anyway. I had learned from the Seris to mind my own business. It was easy to rationalize that they had not yet attained enough exposure to the outside world to be fully effective. But every time I stood back and looked at what the situation was just a few years earlier, I saw enormous changes. The Seris were still there and their numbers were steadily growing. Things worked out and the Seris tended to come out on top.

The Seri population had sunk to fewer than two hundred in the 1920s and 1930s but had more than doubled even before I began my studies there. As I mentioned before, a number of earlier reports tended to treat the people in a negative manner and contained substantial misinformation. For example, WJ McGee's late nineteenth-century tome is full of blather and racist prejudice although it also contains useful information, especially the photographs and other illustrations.

In the 1930s Dane and Mary Coolidge wrote *The Last of the Seris*. They stayed at the ramshackle American sports lodge at Bahía de Kino, which was largely an escape from the Prohibition era in the United States. The Coolidges' visit was still a part of oral history when I began working among the Seris. Ed and Becky Moser learned that the Seri consultant whom Dane Coolidge worked with told things to another Seri who spoke Spanish, and he related the information to a Mexican who translated it into English for the Coolidges. The Coolidges then put together their book on the "exotic" Seris. According to Seri oral history, they would get together and talk about what would be fun and interesting to tell these gringos. The amazing thing is how much of the book is right on, but it's also interesting to see what the Seris invented. In contrast, two other early works stand out for their honest observations—A. L. Kroeber's 1931 report based on his brief fieldwork in 1930 and Charles Sheldon's journal from his December 1, 1921, to January 10, 1922, trip.

It was an accident of history and geography that so much of the environment and traditional culture was still extant in the latter part of the twentieth century. Mostly it was the lack of fresh water and the hostilities that went right through the early 1930s and smoldered for decades that kept most everyone else away. In the wake of President Porfirio Díaz's regime and the hardships of the Great Depression, racism was in, political correctness had not been invented, and minorities in Sonora such as the Chinese and Indians, including the Seris, suffered unspeakable prejudices.

The paucity of fresh water created a unique situation. One summer in the early 1980s, there was a time when there was no water in Punta Chueca, a village on the desert coast opposite Isla Tiburón. A Mexican truck driver was hauling water from Kino. He charged some Seris the equivalent of US$10 for a drum of water and laughed about how he ripped off the Indians. But the water-starved Seris thought that was downright cheap. One of the most poignant Seri song-poems has the line "We walked all day and the waterhole was dry." It was a calamity if a man carrying drinking water back to camp stumbled and fell and broke the clay water vessels. Once I found a waterhole just by following a trail littered with pottery sherds. The advent of five-gallon metal cans in the early part of the twentieth century was a great boon. Hiking about on Isla Tiburón, the Sierra Pinacate, and elsewhere along the Sonora coast, but especially on Tiburón, I often thought that the greatest luxury would be to have all the water I wanted to drink.

It really surprised me to find out that many of the well-known life-sustaining waterholes were so dinky. I read about the famous waterhole at Tecomate at the north end of Tiburón. The whole bay is called Agua Dulce, as is the whole wide central valley, which is perhaps twenty kilometers long. The waterhole is shown on maps, even the older ones. The water was so dependable, and right at the shore, that the Seri camp there was essentially permanent. When the Seris told us about plants on the north end of island, it was often in terms of the distance or direction from Tecomate. So when Ike said it's easy to land there, I was really looking forward to seeing this famous waterhole. We landed at the airfield, one of the better ones, and walked over to the waterhole. We took canteens to fill, came to a broad depression, and walked down to the bottom. I don't know what I was expecting, but here was a little mud puddle covered by a dead pelican.

The late 1960s and 1970s were heady times. As students and young academics, we were busy questioning authority and war, opposed to the stupidity of nuclear weapons, and we thought we could change the world. Native values were being "rediscovered." Sometimes there was an unrealistic utopian worship of anything native, especially Native American. I was at a student gathering where a hipster said to a Navajo friend, "How wonderful to be an American Indian," and she said, "Of course, unless you have them as parents." Sometimes I had to hide the fact that I was working with Native people in order to avoid having to bring along enthusiastic followers. I grew weary of armchair idealists interpreting other people's cultures and even deciding what's good for someone else. The Seris were just like anyone else but with a different background; some were wonderful and fun to be with, others jerks and obnoxious; some were brilliant and creative, and others dullards.

I learned from Tom Sheridan that the ideal in any fast-changing culture is to try to keep the best from the old while taking what you want from the new.

But sometimes it didn't work that way. Even so, the hunter-gatherer, water-limited traditional culture of the Seris provided the people with an interesting background for dealing with their fast-changing world. In traditional times it was water that determined how many people could stay in a place and when they would have to move on. Often it was only weeks or even less before a waterhole was exhausted. Sometimes it would be months. I think people also moved when a place started to stink. Just think, you have endless clean sandy beaches to camp on. You can live by the seashore, move whenever you want, and have all the fresh seafood you want. More than once we would be out in the desert or at sea in wonderful weather, in paradise really, and someone is showing how a certain plant is gathered, or we are hunting sea turtles. And so often someone would say, "We were so strong and healthy then; the food was healthy." (There was no diabetes when I started my fieldwork, or none that I knew about.)

But there were downsides to the old way of life, and sometimes there probably were food shortages. On a number of occasions I asked, "Would you want to go back to those times?" "Certainly not! Those were hard times!" When one got too old to keep up with the group when it was time to move camp, the elder was left behind with some water and a dog.

Outside diseases could be devastating and were greatly feared. In the nineteenth century when people returned to their coastal homeland after being in Hermosillo or in contact with the outside world, they would fumigate their huts with the black oily smoke from burning *Koeberlinia* wood. Until recently there was little resistance to measles, which from time to time had taken a terrible toll. Once, in the early 1970s, I was in Punta Chueca and there was a boy who developed a skin problem. It was feared that he had measles. The news spread like wildfire, and within minutes all the Seris except the stricken boy's family were packing and fleeing. There was terror in the air and in less than fifteen minutes every Seri boat and pickup truck had left; every Seri family was gone except the boy's family. I managed to get medical help for him. It was not measles and his problem was readily cured.

People so limited by drinking water had to discourage gatherings, and neighbors could not always be welcome. There was not a strong reward for cooperation that would foster people gathering together. And even in traditional culture a price was paid for any service. Even a child leading a blind man on a stick was given some small token. So when we went somewhere, did something, when people spent time, I was expected to pay something. When

I came to Desemboque or Punta Chueca, there were things I was expected to bring—yard goods, certain tools, and the men always admonished me to bring .22 caliber cartridges. Guns were really illegal. The men kept their old rifles carefully wrapped up and buried in the sand in their huts.

The Seris abhorred stinginess and even traditionally it seemed that no gift or price paid was ever enough. They would laugh about it, but instead of thank you, it was "*xo*," which means "but." Once I brought Chico Romero, then old and blind, a huge expensive canvas tarp to cover his house. It cost me a lot. His response was "*Xo*, I need a knife [also]." I suppose if I am asked to give one characteristic of my experience with the Seris, it is that they were always direct. In their traditional culture, there was no word for hello or goodbye; you were either present or not. And please and thank you were equally superfluous.

If someone specifically asked me to bring something, and even if they would offer to pay me what it might cost, I learned not to do it unless I was prepared to bring the item as a gift. (But when I brought a gift, I would get something in return, such as shell necklaces, which had become popular items for sale to tourists.) I was obligated to turn over the goods but getting paid was sometimes a different matter. There was an amusing attitude about overdue payments, which were called *cuentos viejos*, and "nobody pays old bills." So while the acculturation was sometimes difficult, the Seris were fast learners.

Various aspects of social organization commonplace among other cultures were often not readily apparent among the Seris. People of such a small and dispersed population had no need of political structure except a war captain when necessary. The Seris hated authority. You were supposed to mind your own business. You didn't tell someone else what to do. There were strict speaking taboos; for example, a son was not supposed to speak directly to his father or father-in-law, and just think how much conflict that can avoid.

Sometimes people would tell me that the Seris really did not have any religion, probably because traditional Seri religious beliefs were not organized as a visible church and were not known or understood. Traditional beliefs and practices were intertwined; they did not exist in isolation. Vision quests, creation stories, various aspects of oral history, face painting, song-poetry, and much more were variously intertwined with religion.

Most of the traditional religion did not function after the people settled in villages. The old social systems were breaking down and sometimes there were unpleasant incidents from drunkenness. For many reasons the Catholic church did not appeal to the Seris in modern times, and there were cultural, historical, and logistic reasons the Seris were not a practical objective for the Catholic church from Hermosillo. But the rather conservative Apostolic Church, which began missionary work among the Seris in 1953, garnered appeal.

Armchair academics often condemned the arrival of the Christian religion. My non-involved observation was that it filled a gap and helped curb some serious drinking problems. Much of the service included singing, so church members came to be called the Singing Ones. I remember observing the service one hot summer night. In the middle of the sermon, someone got up to stomp on a scorpion crossing the dirt floor.

When we sat down to eat at the heavy wooden table in the cement floor dining-kitchen-everything room in the Moser's beach-damp house, some of the Seris pressed around the windows staring at us, watching us eat. At first it was kind of unnerving. One American visitor said, "Those poor people, they are so hungry," but in reality we were the best and only show in town. Becky prepared meals for some of the old people. Once, by the back door, old man Burgos grabbed my arm and said he was hungry and had not eaten in three days. He had just finished his breakfast of eggs, toast, and coffee at the front door. But if I confronted him, he would just laugh. Becky once told me that if she had enough money she would open up a seniors home in Desemboque.

Old, infirm, and blind Catarina lived in a little hovel in front of the Mosers' house. Becky fixed meals for her, cleaned her up, brought and emptied her plastic toilet bags. The poverty and troubles were sometimes heart wrenching, which stimulated tourists to bring piles of used clothes. Much of the time the Seris knew these were just old clothes and tossed them out. There was a lot of dignity and laughing and fun.

Sometimes I would tag along on a sea turtle hunt. As I said, it really bothered me to see the turtle populations crashing. And I got involved with Kim Cliffton, Dennis Cornejo, and Phil Regal in a broader view of sea turtle biology and conservation. During the summer the men often hunted turtles at night, sometimes following a phosphorescent trail in the water. They had great knowledge of sea turtles. The men were proud of the way they could turn their wooden boats in tight circles following a turtle. The boatman responded to a complex series of signals from the hunter standing on the prow. The harpoon was thrown with great accuracy. Adrenaline surged as a great beast was harpooned and hauled over the gunwale. The excitement was thrilling, and I wanted them to be successful even though I felt sorrow for these declining animals.

One night after harpooning several big turtles, we beached the boat on the east shore of Tiburón. The men wanted to rest and smoke cigarettes. It was a hot humid summer night, and I started to go up the sandy bank. One of the guys grabbed me and warned me to stay on the open beach, that the dunes are full of sidewinders. That was a turning point. No Seri had ever expressed a concern for my well-being. Little things like that came up from time to time, reminding me that they cared but just didn't talk about such things.

When I started my work with the Seris I was still a student and a biologist very much influenced by academic tradition. Anthropology in those days was suffering from accusations of not being scientific. So I tried to fit the Seri information into a more standardized system of data gathering and devised an academically suitable information form. One afternoon five Seris piled into my four-wheel-drive vehicle and we drove out of Desemboque and up the dry Río San Ignacio canyon bottom past Pozo Coyote. The palo verde trees were in blazing yellow flower, and the women were filling buckets with juicy orange berries from lycium shrubs. It was a magnificent spring day in the desert.

The usual method of operation was for the Seris to show us plants they thought we should know about and then show us how the plants were used, and so forth. Or sometimes there were things they wanted to collect; we would take them where they wanted to go and observe and record the action. But this time I got out my clipboard and started systematically running down my standardized questionnaire. The Seris at first tried to accommodate me, but then Rosa Flores turned to me and said, "Do you want to do this your way or our way?" I threw away my questionnaire and went back to pencil and paper. I once tried a tape recorder, but I would get stuff recorded and have no way to check it. With a pencil and paper, if there was something I did not understand, I would ask again until I understood what was going on. I filled multitudes of pocket notebooks.

I did a lot of fieldwork with Jesús Morales. He was a kindly and tough little old man but had a reputation of being rather wild in his youth and having killed a few men. He was leathery, wrinkled, and skinny, and wore oversized dark glasses. His knowledge was vast; he knew songs and history unknown to

Richard, notebook in hand, listens intently as a Seri woman shares her knowledge of native plants. Richard Felger archive.

Jesús Morales at home in Desemboque, January 1967. Photo by Thomas Bowen.

anyone else still living. His knowledge was greatly respected among the Seris. We worked well together. One day in November 1969, he promised to show me some important places and plants on Tiburón. I would pay for the boat trip to the island and also pay for a young Mexican to help carry water and camping gear. I showed up at Punta Chueca at the agreed-on time. Jesús was not ready. I waited and finally asked when, and he just said "soon." There was not much else to do except wait for him to get ready. I went away and came back many times; it would have been too rude to sit there and wait. The next morning, same thing. Three days later he was ready.

I had a hard time dealing with the three-day wait, but it was worth it. On the island I filled one notebook after the other. I carried a full backpack, and Jesús and the Mexican helper each carried a *palanca* (carrying yoke) with plastic water bottles, sacks of food, a coffee pot, and blankets suspended from each end. We crossed the long bajada on the east side of the island and up into canyons at the base of the mountains. One night we slept on a solid rock ledge by a waterhole. Jesús cut two large bundles of cane, or carrizo, to make model balsas (reed boats or rafts) to sell and showed me the intricacies of the process. That was one of my most productive field trips.

We had long heard about a certain bean, called ***haap***, from the middle of Isla Tiburón. The beans were one of the major harvests, but the region seemed too arid for any ordinary bean. There was a campsite, called Haap Caaizi Quih Yaii, or "Haap-Bean Users' Place," and I made several futile trips to the island to find it. Once I even went to the right place, but the Seri man who took me there could not locate the plant; it was too dry. For the final trip to locate the elusive bean, on December 11, 1976, Rosa Flores was our guide. Rosa, Cathy Moser, and I climbed into Ike Russell's airplane in Desemboque and went to our rough little landing place near the middle of the island. It was unusually hot at midday, but Rosa wore the jacket I had given Jesús Morales the previous winter, which had somehow ended up in her possession. She wore a long full skirt and flip-flops and headed straight east to the haap bean site. The rest of us had a hard time keeping up with her as she marched between the creosote bushes, mesquites, ironwoods, and palo verdes—she did not stop until reaching the site several hours later. She had not been to this place for more than thirty-four years, not since the early 1940s when she was about twenty-five years old.

Cathy Moser, Rosa Flores, and Richard at the tepary bean location in the central valley of Isla Tiburón, December 1976. Richard Felger archive.

The place where this tepary grows is a huge, nearly barren, north-facing hill slope of large rough black lava rocks—really just a big rock pile. The beans grow up through the rocks, which seem to act like giant mulch. We found only part of one vine, one dried stem with a few shriveled leaves, and one dry intact pod. It was a wild tepary bean, no doubt about it, and the descriptions we had been given were indeed accurate. It was just that no tepary had ever been documented from such an arid habitat, so I was uncertain about what to expect. We were about two weeks too late; all the other tepary vines had dried up.

I knew that if I even touched the pod without firmly holding it together that it would explosively separate—the valves would elastically and violently twist apart, flinging out the seeds with no hope of recovering them. Before I picked that one intact pod, I reached over, pinched it together, and carefully placed it in a small brown paper seed envelope. When I let go, it went "pop." There were five small gray and blackish mottled seeds. I brought them back to Tucson and grew them, eventually getting a good seed stock. The next year I brought Rosa a full bag of haap beans and also gave a seed stock to Native Seeds/SEARCH to grow and distribute.

One of the National Science Foundation grants included funds to study historic museum collections. I scoured photo archives and collections at the University of Arizona and various museums and libraries in California, the Smithsonian Institution, the Heye Foundation, the Museo de Antropología in Mexico City, and the Amerind Foundation. Most of the artifacts were from the early to mid-twentieth century, and the Smithsonian had artifacts and photographs from McGee's 1895 expedition. Having once worked as a curator at the Los Angeles County Museum of Natural History, I knew my way around museum collections.

I was generally welcomed at the various museums and was able to identify artifacts and the plants and animals involved, provide uses, and correct misinformation. Everyday items, sometimes stuff that Seris had obviously thrown away as trash, were now labeled and curated. Usually I had to sign forms and put on white cotton gloves to touch trash that was now an artifact.

Much of Edward Davis's collections from the 1920s were at the Heye Foundation in New York. Davis was a rather flamboyant collector of American

Indian arts and artifacts. He visited the Seris in the 1920s and made a significant collection. He bought a one-man turtle-hunting balsa, apparently the last one and the one he photographed. Some of the older Seris remembered or knew of Davis. Roberto Herrera told me that I could see that balsa "at the university in New York."

The Davis collections were in the Heye Foundation's warehouse in a remote industrial area. I got there with the help of a New York native as a guide. We rang the bell by a little door in one corner of a gigantic warehouse. A guy with a long beard answered and let us in. In the dim light I stood in awe of a dazzling vastness of Amerindian treasure. I didn't find the balsa but I located a pile of Davis's artifacts.

WJ McGee visited the Seri lands in 1895 but the people fled their camps before he could approach them. He "collected" a large Seri balsa, hauled it in a wagon across the desert, shipped it to the Smithsonian Institution, and had it photographed sitting on the lawn in front of the museum. I was able to study McGee's collections, to actually touch links to the Seris in the nineteenth century, to see physical objects that had been vividly described by Seri elders. The balsa, nearly 30 ft (9 m) long, sat on the top rung of a series of catwalks in the attic of one of the Smithsonian museums. Deep piles of silent gray dust covered everything like moon dust and poofed out at each step and rolled back over my shoes. There were all sorts of boats and other oversize artifacts. None seemed newer than the nineteenth century.

Over the years the balsa had collapsed like a deflated whale. I knew it was formed of three huge tight bundles and could picture what it once looked like. I examined the frayed, broken cordage that had once held it together, and it indeed looked like it was woven from mesquite root fiber. When I wasn't being watched I snitched a few rope fragments that had fallen on the floor, and back in Tucson I examined them under a microscope to confirm that they were mesquite root.

Descending from the dust palace, I was shown the cabinets with Seri artifacts. There was a *Jacquinia*-flower necklace also picked up by McGee. The Seris liked necklaces made from the dried orange flowers because whenever you put them in water, the flowers swelled up, the orange color brightened, and they looked fresh and lifelike. I was curious to see if the flowers would still perform. I was sure that nobody would readily let me try it without a lot of red tape, so I asked where I could get a glass of water because I had a sore throat. I got my glass of water, and the sore throat threat was enough to be left alone long enough to dunk part of the necklace in water; the flowers dutifully swelled up and looked as fresh as the day they were picked.

Our ethnobotany book project was just plain open ended. Every sentence, every paragraph brought up another question, and there were people who could provide answers. So I often went to Desemboque or Punta Chueca with specific questions and those opened more questions and vast knowledge. We were constantly adding new information. I got so tired of being asked when the ethnobotany book would be done that I would say "soon." While visiting the herbarium and botanical museum at Harvard, I met Margaret Towle, who wrote *The Ethnobotany of Pre-Columbian Peru*, a work that I held in awe. I was lamenting that it was now nearly twenty years and our book was still not published. Dr. Towle replied, "Only twenty years? My book took thirty-five years."

When I set about to organize the thousands of jumbled pieces of information, I searched about for a model ethnobotany. But I didn't find one that suited the information we had. So one day I sat down and tried to make an outline, discussed it with Becky Moser, and after many drafts, we finalized a workable format. It all seems so matter-of-fact now, but at the time I felt a real dilemma about how to organize and present the diverse information.

Finally, one day I said, no more field trips to work with the Seris, no more entries, no more information. Whatever we've got, that's it. If something new shows up. I will ignore it. Of course that's like telling a kid to look at the candy but don't eat it. I worked on a section, gave it to Becky, and she worked on it; we worked on it together; we checked with Seri elders; I sent parts out for review to colleagues; and piece by piece it came together. If I came to an unresolved question, I tried to force myself to erase the offending text. Before I left the Desert Museum, I spent almost half my annual salary on a gigantic word-processing computer: I got the expanded memory of 90K and a service policy because it kept breaking down; it had eight-inch floppy disks, and each file held about two pages. With that monster, the book was finished.

When we submitted the first proposal for the Seri ethnobotany book to the University of Arizona Press, the director sent us his usual eighteen-page single-spaced letter that basically said that we should remove as many illustrations as possible, and then look through the few final selections and wheedle those down in number. Then Steve Cox took over as the director, and most of the illustrations remained. And then started the long process of working with a copyeditor. Sometimes I had to fight to retain the most important images and concepts. I seldom gave in and am glad that I held to my convictions. But there is one thing that I regret giving in to: Just as the book was going to press I let the editors remove a carefully worded paragraph in the preface that said that this book is only a fragment of the traditional knowledge, and we encourage further work to record and preserve traditional Seri knowledge. But the editors said that our book was a definitive work and wanted the paragraph removed.

We knew we had only scratched the surface. So many times, and this is one of the things that kept me going even during some of the trying times, a Seri elder would say we needed to write down this information. They knew most of it would go with them, and that the young people were not interested. But in reality the full span of the traditional way of life was no longer feasible. The young people did not have the luxury of activities that were not practical. The economic situation was hard.

The University of Arizona Press did a magnificent job on the production. Harrison Shaeffer's design is elegant. But when the book first came out at $64 in 1985, I was devastated by the high price. Would I be an author of a rare book? We had to give away copies, and the Seris loved the photographs, especially of relatives and friends, many of them deceased.

I was beseeched for freebie copies, and even at author's discount we were not able to hand them out like hotcakes. I tried to explain my predicament to some of the Seris, and one guy said, "Look, Ricardo, don't tell me that you are not making a big profit on an expensive book like that. We know you can't be that stupid." Then I got trounced by some do-good idealists who chided me for not publishing the book in both Seri and Spanish. Yikes, it took over twenty years to do it in English. I told them that was a wonderful idea and asked if they would finance the translation and support the publication. They accused me of being a capitalist pig. We would love to have the Seri ethnobotany book published in Spanish, and this is an open invitation for anyone who wants to take it on. Some translators started but did not carry through.

Times and conditions have changed, new challenges have arisen, and the Comcaac themselves have been recording and interpreting their heritage.

One spring I worked with Tita Braniff on arranging a conference on cultural and biological investigations in Sonora. It was a long conference, held in Hermosillo. Just about everyone doing cultural and biological research in Sonora was there. We ate a lot of good food and had a great time. After it was over, Mahina and I rode with Ed and Becky from Hermosillo to Desemboque. We arrived in Desemboque late in the day and got out of the car to the familiar scene of the sparkling sea beyond the garbage and plastitrash blowing all over the place. Becky exclaimed how beautiful it was to be home.

3

# CROCODILE SMILES

THE SUN WAS barely up and already my friend and I had found several new plant records. It was exciting to find *otate* bamboo and other tropical grasses and perennials on this north-facing slope above the water. Things I thought had been gone for a hundred years, gone when the great flood of the Río Fuerte, in northwestern Sinaloa, washed out the cienega, cut the river channel below its floodplain, and left the acequia-watered fields and orchards high and dry. Real estate values plummeted overnight, the town of Ahome was relocated downstream, and then they started grazing and trampling and woodcutting and overused water all over again. But back then there was still endless gallery forests of giant cottonwoods, willows, Mexican bald cypress, and tropical figs overhanging the clear deep dark slow water.

I started moving a little faster after I looked up and saw the sun would soon poke through the yellow-red mackerel sky and turn the day sweaty hot. I slipped on the steep grassy river embankment below the bridge, wet from last night's thunderstorm, and nearly stepped on something that looked like an adult alligator, but it couldn't have been because there are no alligators on the Pacific Coast, only in the southeastern corners of America and China. I scrambled back up the grassy slope, wiped off more sweat, and looked back down on the largest, fattest, most magnificent crocodile I had ever seen. Green oscillations on the olive green-brown body and the permanent smile you see on all twenty-nine crocodilian species.

When I was a student, I saw young crocodiles sunning on fleshy salt marsh plants among the mangroves below Ahome at the delta of the still-untamed river. The evil professor and I spent three days trying to get across the flooded river in the first days of the monsoon. The water had gone down enough to get the old Buick onto the hastily built cable raft that was ferrying cars across the river. But first we had to get a tractor to pull the car out of the ditch at the side of the so-called road, which, when we came here three days earlier, was a slick sliding expanse of wet mud. No matter how slowly we crept along, the car would lose its traction and slide sideways into the ditch.

We ate bananas and drank beer because there was nothing else to do, and it rained again. Ninety-foot buttress-rooted *Ficus* trees and tropical cottonwoods bobbled along in the frothy chocolate water, and by and by the bridge went as they always do. People were paying $100 to put their cars up on big trucks that plowed their way across the leaning bridge because it was too far under water for passenger cars. People were talking about all the *mordidas* for another big construction when the bridge would be rebuilt, and as always, it would be almost big and strong enough.

The cars and trucks that managed to get across never yielded a right of way, and like a thousand fat pigs in the mud they were gridlocked in the magnificent monsoon. So we decided to try a new road and ended up in this toad pond with horny *Bufo* males scream-croaking all night, their vocal-chord pouches blown out like bubblegum. It was so loud we had to yell in each other's ears and finally just gave up and drank more beer and made our contributions to their breeding pond.

It was during those three days mired in the mud that I first learned of Río Fuerte crocodiles and went down the river looking for them. It was already steaming hot in the early morning when I first saw those little crocs. As we crawled through the stubby mangroves to get close enough to photograph them, we swatted mosquitoes and cursed *jejenes*, those fierce little flies that burrow under your skin and then you see them and feel them for days. Under the microscope they are all proboscis with jagged ripsaw teeth.

I guess it was the *jejenes* that drove the evil professor back to his lab at the brick university while I was assigned to carry on the all-important fieldwork. I had the buckshot pistol that he left with me to shoot lizards for his research. But I decided no more herpetology since I could no longer stand to put my brothers in formaldehyde. My local guide and friend Pancho, who did not share my reverence for the northernmost crocodiles, wanted to know if the gun would kill big *cocodrillos* and coyotes. I said, "Of course not," and was glad I didn't have heavier arms. Just before I left, he stole the professor's gun.

I had heard that in the early twentieth century, a huge crocodile had been shot in Guaymas harbor and taken to a museum in Mexico City, and that there were crocodiles in the deltas of the Río Yaqui and Río Mayo. The Yaquis tell of crocodile women coming ashore, peeling off their hides, and placing them in neat bundles by the beach. After going to the villages, seducing and marrying young men, they went back to the beach, put their hides back on, and swam away.

In the 1940s Gwyneth Harrington and her friends visited the Seris on the vast desert coast of Sonora north of Bahía de Kino. She took paper and colored pencils, and Aurora López drew a sea creature, a crocodile with a fixed smile and a thick belly and a curved tail just like the huge crocodile I was now looking down on. I saw the drawings in acid-free plastic covers in a drawer at the Arizona State Museum Library, and the crocodile drawing riveted my attention. Gwyneth labeled it a "mythical sea creature." But it didn't seem mythical; it had key characters of a crocodile, or *caimán*, as they are called in Latin America. In another drawing, the back legs were missing, like a Yaqui mermaid.

I kept staring down from the bridge at the huge crocodile, its inner tube–like belly half inflated on the wet mud, and well-fed puffy hands and feet spread-eagle absorbing the warmth. I was surprised to see my friend sitting near the water under a large seep willow that filled the air with early morning streamside aroma. She was hanging out with the crocodile; or rather it looked as if she had been hanging out with it for a long time. I don't mean to say that they talked to each other or other such nonsense, but she was just calmly sitting next to it. I tried to warn her—she must be a bit daffy not to realize it is a crocodile, not an alligator. She said not to worry.

Large crocodiles are more dangerous than alligators, and even large alligators can be dangerous, though they are easier to gauge. I raised three baby alligators at my childhood home in L.A. Once I tried raising a baby crocodile, but it never ceased being mean, so I found it another home. For many years it lived with my friend's old yoga teacher, until he died. Did the crocodile eat him?

# 4

# CANYON INTERLUDE

FOR A FEW YEARS I worked in Tucson at the Arizona-Sonora Desert Museum, established the Research Department, and kept active in international conservation. Set in a majestic saguaro desert, it was where people wanted to work, or fantasized it was where they wanted to be. More than one director had his hand in the cookie jar and fired people for petty reasons. Probably just like elsewhere, where there is never enough money except for top dog administrators. Eventually I realized that most museums, botanic gardens, zoos, and such places cared little for their employees, and working at such places was too often a poor idea if you wanted to be creative and treated with decency. The best part of the Desert Museum was the opportunity to do interesting things away from the institution. That was when I still wanted to go to faraway places. Before I realized I don't want to leave home place.

There was a special events trip for museum members to the Grand Canyon in early fall. I was interpretive guide, the expert. I knew the plants and critters though I had not been down that canyon. Almost a week floating the river.

Our quiet passage down the Grand Canyon was like seeing new friends pass by in reverse, peeling time as we progressed downriver through geological time to older and older layer-cake cliffs. I imagined Mesozoic swamps and plants and creatures in reverse evolution fast-forwarding to the desert plants now growing out of cliff cracks and canyon bottom: Fremont cottonwoods, coyote willows, and aromatic seep willows. Three big rubber rafts tied together bobbling down the tamed canyon river. Slow good times with nice people and careful lovemaking when everyone else was supposed to be asleep. The river was low because there wasn't much water being let out of Glen Canyon Dam—until the last day, when water surged.

Any mention of Glen Canyon brings me back to the drowning of that special place. Phoenix and Las Vegas, satiating their thirst for water and power, moving mountains of concrete and steel to plug the canyon. When I was twenty the evil professor sent me with another student to document the plants, reptiles, and amphibians before they drowned. I started collecting plant specimens with enthusiasm but puked and fled when we encountered brown ooze creeping upcanyon. You don't see the connections between flipping light switches and the electric towers above the rising dam. I suppose for most people it's like world wars before you were born, remote and unfelt. You don't have much empathy for animals wiped out before you were born. Who cares that our ancestors wiped out the last mammoth, that Vitus Bering's men slaughtered the last greatest biggest gentlest sea cows? Who cares that Glen Canyon is dead under Lake Powell (we called it Lake Bowel), marble canyons beneath sludge? Someday the dam will crumble, and floodwaters will cut the Grand Canyon deeper, and the Colorado River delta will return to an epicenter of life.

Grand Canyon was 100 degrees in the shade, but in and out of the water made for good times. From the boat landing I drove back to Tucson, and it seemed like I'd been gone only a day.

# 5

# CLIFFHANGERS

ON THAT DESERT MUSEUM float trip in the Grand Canyon, the boatmen who were supposed to be the leaders of that trip were show-off rock climbers. When we stopped for lunch, a side canyon, or some scenic place, they would daredevil scale a canyon wall like a gecko.

No way I would go up or down a cliff unless it meant some prized specimen. I don't like scrotum-tightening scary places, but I got into a couple anyway. Once when I was young and naïve, the only way I could get newly found specimens of cliff-dwelling orchids and bromeliads was over a canyon cliff on a rope. We were near the entrance to Arroyo Gochico, a narrow gorge at an opening to the Sierra Madre Occidental. It wasn't a big cliff, but the plants were too high to get with a pole. I didn't have a rifle, and shooting them down like botanists do in the tropics would be dangerous in a narrow canyon.

We went around to the top, tied a rope to a small stout tree, and I slid down. It was easy with toe and finger holds. Halfway down, after bagging some specimens, I found myself eye to eye with a pointy-nosed green rattlesnake coiled on a tiny ledge, flicking a shiny black tongue, cat eyes in a black bandit stripe across a scaly face. I suddenly realized what was going on and slid down fast and saved emotions until standing a few feet from the rope. I yelled to Ramón to move the rope, but he couldn't hear because of the silvery water running through the gorge. Some places so narrow you can almost touch both walls. Bright red and green squawking military macaws made fun of me standing with my feet wet and wondering what to do. These deep narrow canyons hold moist air, and the shaded walls afford special niches for tropical plants. Tall slender palms grow from crevices in sheer rock. Rock figs cascade white roots down the walls as if melted, reaching canyon bottom water to grow into large trees, often so top heavy they break away and crash into the canyon. You don't find these plants in the surrounding open dry tropical deciduous forest, scorching hot and dry before the summer monsoon.

I walked back up to where Ramón was smoking a cigarette while the pack burros nibbled dry grama grasses. Later in the day, in the same canyon I saw the same orchids and bromeliads, as well tropical cacti, palms, and ferns, growing within easy reach.

The other memorable cliff event was on Isla San Pedro Nolasco, a steep-sided granitic mountain sticking up from the sea northwest of Guaymas. We had been camping at Bahía de San Pedro before it became a favorite rendezvous for drug runners heading up the Gulf of California. In the early morning we crossed the channel to Nolasco, anchoring at the southeast cove called Cala Guina, meaning chigger cove because is it so tiny and narrow. I climbed the high cliff above the cove, but in my enthusiasm for new specimens, failed to

View down the steep slopes of Isla San Pedro Nolasco to Ike Russell's boat *Ofelia*, anchored in Cala Guina, where Richard went ashore, November 1964. Richard Felger archive.

note my route and couldn't retrace a way back to where Ike and Alice were waiting in the *Ofelia*. They were only tens of feet away but hundreds of feet below.

I yelled to Ike to meet me at the northeast canyon where the way down is not difficult. Only about a mile or two, and I thought it would be an interesting walk. I didn't know it entailed numerous descents and ascents of steep hanging canyons. At one point a rock gave way and I was sliding to a high sea cliff. Instinctively I fast grabbed a big cholla, spines and all, and held on tight and felt intense pain only after realizing my luck. After half an hour inching up spread-eagle over unstable guano-white loose rocks, I finally stood on solid rock.

It was midnight by the time we got to the once-elegant Hotel Casa Grande in Guaymas. After all that, the bedbugs made sleep impossible, so we sat up on metal chairs in the interior garden patio and waited for dawn.

6

# WALKING OUT

THE MULES HAD just been hobbled, and I was tightening the plant press straps, thinking about the day's discoveries. A white-flowered *Odontoglossum*, not only an astounding northern record, but it looked like a new, undescribed species. I even managed to catch one of the iridescent green orchid bees pollinating the newfound orchid.

We were on the Chihuahua-Sonora border, a few days into the Sierra Madre Occidental from the end of the road at Don Nacho's ranchito, and had stumbled onto poppy fields. We were about to have late afternoon coffee when a couple of gun-toting locals rode up and slid off their saddles. Don Nacho gave them coffee and they talked about the weather. Turning to old Nacho, they said, "You can go but we're not going to permit the Señor to leave." That means they're going to kill you.

We offered them more coffee, which they politely drank sitting on their haunches. I was never very good at it but managed macho stoicism sitting on my haunches too. They went off as it got dark, reminding Don Nacho about returning later.

I didn't know what to do; the situation seemed unreal. There were interesting specimens to prepare, but I was too rattled. Don Nacho let the fire go down. I tried to talk to the old man, but he wasn't in a talking mood, and besides, his Spanish wasn't that good, and I only knew a few Guarijio words. He was so steady and unperturbed that it almost calmed me.

It was soon so dark that I dared not move and just sat up against the hard scratchy trunk of a giant-leaved tropical oak. Tough leaves so large you can use the water collected in upturned leaves to wash your face in the morning. The pioneer botanist Howard Scott Gentry said those leaves were handy in a dry camp.

Don Nacho gently touched my shoulder and put his old dry hand over my mouth before I could say a word. In his steady silent way he got me up and led me over to the pine forest just beyond the clearing where we had our camp. I could barely make out his crooked finger urging me on. He hooked a finger into my belt and led me on. I wondered what he wanted to show me, and again, even before I could ask, he covered my mouth with his old scratchy hand.

He was taking us somewhere. I wanted to go back and get my jacket. There was food at our camp, the two burros, and my specimens. My camera too. It was pitch dark, and I kept stumbling. He took us up the mountain, almost straight up and over the rocks, where earlier in the day I had found red-leaved bromeliads and yellow-flowered orchids clinging to north-facing shaded cliffs. I had tried to catch the male orchid bees attacking those orchid flowers, as if the flowers were other male bees invading their territory, a special orchid trick evolved for pollination.

We worked our way around colonies of agaves that more than once bloodied my legs. On the narrow animal path he was using for a trail, I started to fall, but he pulled me hillside, away from the steep abyss. Heavy darkness continued for hours when mercifully the moon rose, and I realized our campsite and the poppy fields were already on the other side of the mountain. I thought we might be going for help or to the home of friends, some safe place.

Tiny orange kerosene lights flickered far below and again he took us up higher and away from valley bottom, echoing dog barks, and warm light. We went hour after hour, one step at a time, up and down steep slopes, away from ranchitos. I was getting cold even though part of me was drenched in sweat.

Still he would not talk to me, or it would be more accurate to say he would not let me talk. Up and over mountains. Don Nacho was not stopping. He did not step on anything or do anything that made a noise. I tried to copy him, but it was a hard act to follow.

Although the quarter moon gave some light, I had a hard time finding a good place to put my foot at almost every step. In the worst places he would again firmly take hold of my belt and lead me like a dumb stubborn mule. It must have been midnight when I stopped to turn for a piss and felt my feet, legs, and whole body quivering as it tried to catch up and rest. He wouldn't even give me five seconds after I finished. We plodded on, that is, I plodded on after him.

In the moonlight I saw we were thousands of feet above the canyon floor. We seemed to go so silently that I could hear every night noise of poorwills calling, bats squeaking, and once in a while a cowbell far off below. Some big bats whooshed musty air against my face.

We went way around well-worn paths. Crossing a stream, I slipped on a slick rock and got wet, and still Don Nacho would not stop.

At stream crossings we slowed only to drink. A carpet of big dry crackly sycamore leaves nested between water-smooth granite boulders and the pungent medicinal aroma of seep willow mixed with cattle smells on the damp canyon soil. Coming around a huge rock, there was a large animal growl.

My feet, head, and hands were numb, and my mind opened to the sky. We seemed to just drift over spinescent slopes and through tree-darkened rock canyons. I thought of nothing as new senses washed through like waves on childhood beaches. The trees and grasses and rocks and cactus and sky and night creatures and old Nacho himself were all in sharp focus. Streams of light and heat seemed to flow from every living thing, including the rocks.

The east sky slowly glowed warm with bird calls. Don Nacho led us up a steep high crumbly slope, a hard climb over sharp rocks, and finally into the dusty floor of a cave smelling of mammals next to a rock seep of cool clear water. We slept past midday and waited for dark. I forgot about talking. There were new colors and sounds and I tried to forget about being hungry.

We walked easy in the black and white night world and came around the backside of the village as the roosters called. Plain adobe brick walls and palm thatch roof, the house open to the warm south, beneath giant quiet sabino trees along the now dry riverbed. We sat on homemade chairs on the hard-packed dirt floor, the seats and backs of interlaced rawhide strips with the fur still on. Doña Marina brought coffee and hot tortillas wrapped in a cloth. We ate dark beans and eggs and drank more coffee. I slept like never before or since.

It was bad enough losing hundreds of specimens and all my field notes and camera, but now I had to pay for both pack burros. We said nothing about what we saw.

Some months later I heard that the federales raided the poppy fields. Howard Scott Gentry told me about his harrowing encounter with poppy growers in almost the same place more than thirty years earlier. Nowadays there is a road to where the poppy fields were once incognito, but it is still an unhealthy place, noted for armed robbery. Not far from where the locals say, if you go to Huirocoba, bring your gun, and if go to Choquincahui, bring your coffin.

# 7

# DESERT KAYAKING

WHEN I WAS IN PHOENIX a few days ago, Gil Gillenwater and his brothers were talking about river trips and I said I would like to go along sometime. When Gil called and asked if I wanted to go on a kayak trip down the Gila River, I was sort of surprised. I always say I want to go on an adventure or trip, but most of the time it's just talk. Gil said we would meet on Sunday, January 31, at 8 a.m. on the west side of the Gila crossing on I-10 out of Phoenix, past the Casa Grande turnoff.

Gil said bring change of dry clothes, gloves (bicycle, sailing, half gloves), bicycle helmet, coat, high-tops, and wet suit. Wet suit (don't have), high-tops (only have low-top tennis shoes), coat (have), fingerless gloves (mine are wool for typing when it's cold), bicycle helmet (don't have), a dry change of clothes (I have).

I went to a bike store to buy gloves, and the bike nerd said bicycle gloves are no good if they get wet. I bought the cheapest ones and they worked fine. I borrowed a Farmer John–style wet suit from Jeff, who said there are two kinds of people who wear wet suits: those who pee in their wet suits and admit it, and those who pee in their wet suits and don't admit it.

When I told some friends what I was about to do, they shook their heads and emitted warning signals. Or just plain chuckled when I admitted I had never been in a kayak.

Twice in early January I planned to go to Phoenix, but the 1993 winter storms coming in one after the other made me postpone. Arizona winter rain usually doesn't stop me from highway driving, but we were having an El Niño year with unending rains, floods, and roaring rivers. If you are almost anywhere but the Sonoran Desert, the idea of canoeing or kayaking down a river doesn't sound all that special. But our rivers are dry, bone dry; people ride horses down the rivers and the four-wheelers smash down the remaining vegetation. Everybody knows our rivers are dry and dusty, at least most of the time.

Friend Barbara laughed because it was supposed to rain all day Sunday. When mentioning this factoid, Gil said, "So what? We'll be wet anyway." I knew I couldn't be late, so the night before I put my gear in the car. Left Tucson at 6 a.m. even though 6:30 would have been time enough. I kept watching the eastern sky to see what the clouds were doing. Sunrise was slow because there was a mackerel sky that turned bright yellow and then pink and then gray, the kind of clouds that portend rain. Made a coffee stop and got to the Gila bridge a little before eight.

I park alongside the highway, just northwest of the bridge. The river fills the whole floodplain. It spills out across the desert, filling backwater expanses, stranding palo verde trees. Trucks roar past; my car shakes every time one goes by. I keep looking from the river to the highway, from the raging river to the roaring trucks. There is more energy and mass going down the river.

Gil is only a little bit late. He is with Highly Falkner and several other friends. The back end of Highly's old GM carryall is filled with well-organized gear. I get my jumble of things and climb into Highly's car. I am nervous about leaving my car along the highway, but Highly says the cops will put a sticker on it but can't do anything to it if we come back for it today. I wonder about the "if." Highly cuts across a well-worn track in the median, goes eastward for about ten miles, and turns off northward on the Florence Road (AZ State Route 87). Suddenly we are off the roaring highway and the day starts to be nice. The sky is clearing and air clean, brilliant, and cool. We make a fast pit stop. The ground is damp, the desert spotted with sprouting spring wildflowers. The filaree is starting to flower, and Old World red brome (*Bromus rubens*), hairy mustard (*Brassica tournefortii*), and Mediterranean grass (*Schismus arabicus*) are already smothering the slower-starting native wildflowers. As I zip up, I remember that we are near Sacaton, where Kearney and Peebles worked on *Arizona Flora* before World War II. Where the soil conservation service was busy introducing Old World grasses to improve range lands. Purposely introducing

African grasses that turned into hellfire, the place where the Africanization of the Sonoran Desert got going big time. I am thinking about the first specimens of *Schismus*. In 1926 it was one of the first non-native grasses around here.

The river is over a mile wide. After I get into my wet suit, I realize I shouldn't have been guzzling coffee for the past few hours, but the night before, I had stayed up late reading Amadeo Rea's *Once a River* to get some idea of what the scene used to be like along this part of the Gila. We are in three kayaks. I am in one with Gil. One place the river goes under a low bridge; it's low because we are up near the top. I look up; people above are looking down and wave, and I wave back. Gil yells to pay attention and keep away from a whirlpool with a moving pile of whole mesquite trees, willows, cottonwoods, and unidentified cultivated trees, paddling hard and then fast forward down the river. We are kayaking around saguaros and palo verdes, the river so swollen. In one little palo verde tree, a big skinny diamondback rattlesnake is loosely wrapped around leafless branches, like it's been hanging there for a long time. I tell Gil we need to rescue that poor little snake. Gil believes me and paddles as fast as possible.

Richard in the bow and Gil Gillenwater in the stern, paddling their way down the flooded silt-laden Gila River, 1993. Courtesy of Gil Gillenwater. Richard Felger archive.

Lunch and pit stop on a newly eroded sandy bank. A perfect desert day.

Thirty-six miles go quick. Just the noise of the water and a sky-pig helicopter that is following us. Back at the highway, cops are waiting for us. Highly informs the main officer, Mehringer on his name-tag bar, there is nothing illegal about kayaking. Cop lectures us that four guys died the day before doing the same thing. With authority Dr. Felger informs Mr. Policeman that those guys were drunk, inexperienced, and paddling at night. We are sober and very experienced (after my first kayak adventure).

I am watching the river off to the left in the valley bottom. Thinking about muddy packets of water making it all the way to the Río Colorado delta and out to greet the last lingering vaquitas.

# 8

# SINALOA SHOOTOUT

WHEN WE GOT BACK to Ryan Airfield, west of Tucson, Ike would say, "Made it again."[1] It all seemed so matter-of-fact, spotting plants out the window, flying around mountains, landing just about anywhere. Going to see whales, meeting interesting people.

Eelgrass grows in protected seas along temperate coasts around the world. It is well known from Sonora but factors affecting the southern limit and the situation in Sinaloa were a mystery. A 1910 report mentions a specimen from Altata, due west from Culiacán in Sinaloa, and I want to go there. It is mid-March, and at this time of year there should be detached eelgrass forming rafts at the sea surface. As usual Ike is eager to go—I just have to pay for gas. I ask Kim Cliffton, who works with me on sea turtles (they eat eelgrass), and Tom Backman, who did his graduate work in California on eelgrass, to come along. Four of us, a full plane.

We get to the Russell's early in the morning. Ike is rummaging through metal ammo boxes of nuts and bolts, spare parts, tools, and his medications.

---

1. "Sinaloa Shootout" is from *Backcountry Pilot*, edited by Thomas Bowen © 2002 The Arizona Board of Regents. Slightly revised and reprinted with permission of the University of Arizona Press.

He spends a lot of time shuffling things and packing. I don't question it; he is in charge of airplane matters. Sometimes he also spends a lot of time disguising things hard to get across the border to take to friends in Mexico. Finally, we drive out to Ryan Airfield in the Avra Valley west of Tucson, all jammed in his dull gray 1941 Ford pickup, about thirty minutes from where Ike and Jean live.

Ike has a hankering for old pickups. Once on the way to Ryan, he made a U-turn on the Ajo Highway, opened the door, stuck out his long arm, and scooped up a hubcap. "This is a good one." Like any old rancher, he keeps a lot of spare parts. Once an attorney and I had a rasty little ranch. An old Chevy pickup came with the ranch and I said the pickup is mine, but the attorney didn't give me the title. So, I traded that rattletrap to Ike for a trip to Mexico and he took the truck to Mexico.

We're at Ryan, and Tom probably thinks we are just going to climb into the plane and take off, and Kim has the patience of a turtle. More tinkering. Checking the engine and different parts of the plane. Ike gets into a deep conversation with Fred Grissom, who takes care of the planes in this World War II hangar—wide tongue-and-groove weathered 1940s wood outside walls, and cool and drafty inside like an open cathedral. Fred wears overalls and fusses over the planes like a mother hen—Ike once said he was alive because of Fred. Airplane parts all over the place, his office boxed off inside the hangar, smelling of cigarettes and stale coffee, and girlie calendars advertising tools and engines.

Fred teases Ike with concern about not bringing the plane back again with a green propeller—it turned green the time we went through tall amaranths at the mining claim. Fred once said, "Ike would fly in weather that would drown a duck." And there was the time Ike landed silently, got out, and had his passengers help push the plane to the hangar. Fred said, "Cutting it a little close on gas, aren't you, Ikey?"

Tinkering is a big deal for Ike. He loves to tinker the way I like fieldwork and studying plants in the herbarium. Ike put a 1918 Dodge engine in the *Ofelia*—the ungainly little boat he had built in Guaymas modeled after a fisherman's *panga*—says it can cruise the whole Gulf of California on one tank of gas. It's loud, stinky, hot, and slow but provides lots of tinker time. He tinkers with his pickup. He tinkers with his airplanes. I always carry a paperback book in my back pocket.

Finally, when it's time to load gear into the plane, he scrutinizes everything I brought. He picks up each item and dangles it up and down for weight. Everything is always too heavy. "Do you really need so many plant presses?"

and "Do you have to bring that thing?" There was the time Jean tried to put a sea-worn egg-shaped boulder in the plane—we had picked up the rock on Isla San Esteban and brought it to Bahía de Kino in the *Ofelia*. Then one day I saw that the egg rock had made it to Tucson and sat in a place of honor by the front door. Sometimes, especially in summer, I have terrible arguments with Ike about water. You can't win an argument with Ike for the simple maddening reason that he is always right—except about summertime drinking water. I won't give in on that one—you never forget close calls with water. I learn to bring twice as much water and twice the plant presses I really need so Ike can make me leave half. It's always half.

As I said, you never forget close calls with water. Once on Isla Tiburón, we flew around the east side of the big mountain—the Sierra Kunkaak—until we located the Sopc Hax waterhole, marked by bright green giant cane grass, tucked into a canyon recess at the toe of the mountain. We land at Palo Fierro, and it seems like an easy few miles' walk, but what's easy to spot from the air can be hard to find on the ground and we climb up the wrong canyon. Ike doesn't stop when I stop to collect plants and we get separated. I see smoke off in the distance. He's burned an old packrat nest for a smoke signal and next to it left a note stuck on a creosote bush stick: "I'll meet you at the big rock fig that blocks the canyon." I take a shortcut straight over the ridge to the next canyon and down to the big rock fig blocking the canyon, but no Ike. He must be behind me, coming up the canyon bottom. I wait . . . and wait. Yikes, it's late in the afternoon, so I hightail it out of the canyon and head for the plane. I have very little water—we expected to fill canteens at the waterhole.

Night falls fast. Pitch dark and no flashlight since this was only a day trip. I keep bumping into impenetrable mangroves and try going around but too many cactus and spiny shrubs. I fall asleep on top of sharp rocks and scratchy cryptanthas and wake up with coyote breath in my face. I move, and the coyote runs off but now it's cold and me with only a thin T-shirt. Dawn is way too slow. I am out of water. My tongue is huge, and I can't swallow. I misjudged and have to go several more miles out of my way around the mangroves to get to the landing field. The sun is high, hot, and swirling in a Cézanne sky, only it does not look pretty. The silver sea beckons, but I keep forcing myself to realize it's not drinking water. I know the plane has to be way off to the left of the sea shine and keep forcing myself in that direction. I must be close but can't find the long low flat ridge with the landing strip. I pop a few bursera fruits in my mouth—the Seris say it helps make saliva run when you are out of water.

The first time it almost helps. I climb up into a huge cardón cactus and don't see the plane. It has to be there . . . it has to . . . he wouldn't have left. I keep forcing the steps, and just as I start to walk right through the plane without seeing it, Ike hands me the last tiny bit of water in the half-gallon green glass bottle that's always in the plane. It's not enough, but he is not much better off and has been saving this for me. We climb in and fly wordless across to Bahía de Kino, land at the edge of town and taxi up to Doña Gavina's little home-restaurant next to the beach. We walk in, pick up water pitchers sitting on the tables, and drink, spilling water down us and all over the place. Don't get me wrong; it's not always like that. It's usually fun and easy.

So here we are at Ryan in this huge old hangar crammed full of airplanes. Fred has great talent for fitting the wing of one over the wing of the next one and so forth. We move a few planes out of the way and roll out Ike's Cessna 185 that has a STOL kit (Short Take Off and Landing) and other modifications. I take out my book and sit on an expensive airplane tire.

Finally, we climb in the plane and take off, first for Nogales, Sonora. Always have to land in Mexico and get permits and clearance. I expect delay. If Ike would just give a standard little donation we would be on our way promptly, but he never does. Says if he does it once he would have to do it every time. More to the point, Ike is just plain stingy, stubborn, and likes winning.

The officials know he brings "extra" stuff. They accuse him of flying to Mexico on business, being a commercial pilot paid to fly in Mexico. For such purposes one needs a commercial visa, which means lots of red tape and expense. He comes through too often, in and out, with too many different people. Nobody spends so much time and effort just for fun. The places we go cannot be for pleasure. He has to be up to something. Well, it's not a truce, but there seems to be an understanding and somehow we always get through. In addition, this being Mexico, the airport comandante is responsible if anything happens to a plane that leaves under unsafe conditions. So there is real authority and concern.

The green propeller incident I mentioned a little while ago involves a trip to Ike's mining claim in the Sierra Madre Occidental in western Chihuahua. That time we were clearing Mexican customs in Hermosillo. Ike files the required flight plan from Nogales, Arizona. But before arriving in Hermosillo, he makes an unannounced westward detour and lands way out in the desert. I nearly wreck my back helping him unload an unbelievably heavy wooden box we stash under an ironwood tree. I ask what's in it, and he just says "Dynamite for the mine." We arrive in Hermosillo flight-plan late and things do not go smoothly. Ike won't pay an eight-dollar special fee. The day wears on. I say I will pay it. This is not my idea of fun.

Ike goes to talk to El Comandante for the fourth or fifth time but El Comandante has "gone home for lunch." We get a taxi and go to his home. A woman answers the door and says El Comandante isn't in, although his car is in the carport. Back to the airport. Wait for El Comandante. Pretty soon it's too late in the day to leave. Taxi to a downtown hotel. Ike is a cheapskate unless it involves the airplane. He always picks cheap places. Mercifully this one is not absolutely the cheapest.

Ike worries that some cowboy might find his dynamite—about the only time I notice him being perturbed. This is costing more than eight dollars. Next morning after breakfast, a taxi to the airport. Ike is having a long conversation with El Comandante without me present. We leave and pick up the box in the desert and head across the desert for the Sierra Madre.

We've been at the mining claim for two days and it's time to leave. But it's 4,000 feet elevation and over 100 degrees this afternoon, so we have to wait until morning for cooler (denser) air to take off. Okay with me, I get more time to explore for plants, but Ike has a different plan. We chop amaranth and other weeds that have grown up with the summer rains at each end of the landing field. Ever chop fresh green amaranth as high as your head? Yellow pollen-rain mixes in sweat and runny nose and eyes. The landing field is not much more than several rows of corn Ike pays the owner not to plant. I don't like landing here. There are worse places, but I only go to them once.

We leave soon after sunrise but it's still hot. The plane lifts off too slowly and the prop chops through amaranth just beyond the part we cleared. Barely clearing some scrubby boat-spine acacias and mauto trees, we sort of sink into a vast canyon before the plane can lift up above the mountains.

I will never forget the way Fred shook his head when we got back to Tucson with a green propeller. I didn't say anything to Jean and I doubt Ike did. Ike and Bill Swan had bought the mining claim from Bill's brother-in-law Pat Jenks, who is notorious for treasure-hunting escapades in Sonora. I don't see how they expect to make money on the mining claim, but maybe that's not the issue.

Let me tell you about another dicey summertime event. We arrived in Nogales, Sonora, before midday. Plenty of time to get to Guaymas by early afternoon. It is clear and very hot in the middle of the monsoon season. Long conversations with El Comandante and a long delay. Billowing white cumulus clouds close in and turn to dark nimbus. Lightning, thunder, and torrential rain, and refreshingly cool. El Comandante declares it is unsafe to leave. We wait. I finish my book. The clouds clear and El Comandante almost let us go, but before another long conversation ends the clouds roll in again.

Finally, a break in the clouds, and Ike manages a breakthrough with El Comandante. Ike says, "Let's get out of here." I seldom see him move so quickly. Almost as soon as we are up in the sky the clouds close off the intended route. Ike says, "There's a hole over there; let's head for it." That route closes, and we chase more holes in the clouds. Finally, after chasing lots of sky holes, here we are again smack dab over the Nogales airport. Ike does not want to go back there again. The clouds are high enough, so we fly under them. Ike follows the highway.

There is bright sunlight not too far off at the end of the valley. I look down at Emory oaks on overgrazed hills, and when I look up, the sunlight is gone, the ceiling sinks, and it's dark and stormy. The plane tosses in wind and torrential rain. Ike says we have to land and manages to find a clear stretch of highway. Right away he gets the plane off the blacktop onto a little dirt turnoff at a roadside ranchito. I hop out, open the flimsy wire and pole ranch gate, and Ike drives the plane through. I close the gate and run up to the house for shelter but am already drenched. Ike stops the plane in front of the family gathered under the broad low porch. They offer handshakes and *Mucho gusto en conocerle* (Very pleased to meet you). Homemade wooden chairs with criss-cross leather strapping are placed at the table for us. Dark hot coffee and lots of sugar and pleasant conversation. The storm clears as fast as it came.

We are saying goodbye and thanking our hosts when the head of customs charges into the muddy yard in a chauffeured Thunderbird, surrounded by a roar of police motorcycles. He announces, "We got you this time," and threatens to confiscate the plane. Ike and the head of customs are just beginning their conversation when two more motorcycle police roar up. They whisper something to the head of customs, and without another word they all leave, sirens blasting down the highway.

We thank our hosts again, shake hands all around, promise to come back and see them, drive up to the highway, wait for the traffic to pass, and take off. In just a few minutes we fly over a twin-engine Beechcraft and the Thunderbird next to the highway, surrounded by the motorcycle swarm. Later we learned that they had also made a forced landing and quickly paid a $2,500 fine.

Back to the Sinaloa eelgrass trip. This time we clear Nogales without incident. We head for the coast at Desemboque and hope to find Becky Moser at home. Ed Moser died suddenly last summer, and Becky is carrying on her husband's work.

It is just a few years since Becky and I published our paper on Seri use of eelgrass—the first report of a grain from the sea as a human food resource. *Science* published it and put Helga Teiwes's haunting photograph of a Seri eelgrass doll on the cover. That paper made quite a splash. It was also a turning point for me. I had recently left a generous position at the Natural History Museum in Los Angeles. I had to be back in the desert, out of the smog and off the freeways, but the financial part of quitting was not realistic. The *Science* paper helped in getting grant support for fieldwork, including airplane gas, and launched me into the emerging halophyte-science crowd. Hazel Fontana made the first loaf of bread from the sea using eelgrass flour we bought from the Seris. It was bright green inside.

That *Science* article—as well as a few others—encouraged Becky and me to keep working on our Seri ethnobotany book. The book was still "almost finished" and I felt guilty not having it completed in time for Ed to see. It took only another eight years to finish. Without Ike and Jean Russell it would never have been finished, let alone started.

Ike Russell first brought me to Desemboque. We had been on Isla Tiburón, and I had a plastic bag full of plants. Soon I discovered the Seris living in my study area and realized they knew more than I did about the plants. Those were wonderful times. I was finishing my PhD. Then some academic appointments. We were all so purposeful. Ike had to have a reason to go flying. My reason was that I had my work to do. It made a big difference to have Ike take me in his airplane. It was fun to go places with him, airplane or not. Everybody loved Ike, at least after the waiting. It was even more fun when Jean went along. We landed and camped just about anywhere. And John Dog often went with us.

The first time at Russell's for dinner, John Dog joined us at the table. Ike and Jean's son Luke taught him to gingerly take food off a fork and eat politely from a plate—seemed like he had the best manners of anyone at the table. John's tail stuck out behind the chair. When John got too old to jump into the plane, we lifted him up. He was a big Aussie weighing over ninety pounds. Then after John was gone Weni would go, especially if Jean went. Ike inspired Weni to growl, and Weni inspired Ike to growl, although Weni got along with everyone else. Then my dog would go along; her father was one of the Russells' big white dogs.

As I said, those were wonderful times. We could land just about anywhere. Ike seemed to know every rancher in Sonora or knew someone who knew someone where we would go. We were always welcome, and they were always welcome at the Russells'. I don't know how Jean did it—there was always room for another friend or friend of a friend. Their tree-shaded ranch along Cottonwood Lane was an international institution, like a grand hotel.

There was still no big drug trouble in Sonora. At least we didn't have to know about it or pay attention to it. Once in a while we would see a fast fancy plane on a dry lake bed and some city boys standing around waiting for someone. And only once were we clearly not welcome, when we landed at a forlorn ranch way out in the desert between Nogales and Desemboque.

Back to the Sinaloa eelgrass trip. We buzz the house, and Becky and some other gringos wave, and her daughter Cathy's lazy brown dog Slab is wagging his tail. Seri kids are running for the landing field and everyone is looking up. We land at the long, sandy smooth field near the shore north of Desemboque. Ike pays a Seri, José Astorga, to help keep it clear and had a sign made in Spanish and English that says anyone landing here should pay Astorga a certain nominal fee for his labors. Becky and her guests drive out to pick us up. The Seris press around us. They are not at all meek, always laughing and making fun of everyone, the women all at once trying to sell necklaces, baskets, and ironwood sculptures.

During those years, whenever we flew into Desemboque, the Mosers' house was always open to us. After dinner and the sun went down, the kerosene Aladdin lamps were lit, and it was the most peaceful place I have ever known. Sometimes Ed performed magic tricks. Becky and I worked on the ethnobotany manuscript. There were the day's notes to write up, plants for me to identify, revisions in the text, more questions to be answered. The action centered on the heavy wooden table in the kitchen.

The shower was behind the big silent gas refrigerator. Water was heated on the stove, poured into a bucket with a twist-turn sprinkler at the bottom, and the bucket carried around behind the fridge and hoisted up on a pulley. You closed the curtain and could keep on talking to the people in the kitchen. Ike all too often passed on the shower.

The next few days are occupied with fieldwork along the Canal del Infiernillo opposite Isla Tiburón, botanizing and going places in the desert and along the coast with Seris, recording information on ethnobotany and sea turtles. After dinner, we spread out maps and look at the Sinaloa coast to Altata. The map shows some sandy barrier islands and huge complicated lagoons.

At bedtime we open up aluminum folding cots on the back porch. It is dark, with only stars and the wet sound of the waves and dogs and sometimes donkeys and coyotes. Everyone else is asleep, but just as I drift off another bunch of mangy Desemboque dogs starts barking and yowling from yet another direction . . . then the silent torture waiting for it to start up again, over and over. When Cathy Moser was in college, Ed sent her a double-sided tape of "Desemboque Sounds at Night." She thought the dogs and donkeys were wonderful.

Ike is in the little room off the kitchen—the dampness outside bothers him tonight. He is always coughing a bit. He has chronic health problems, which is why he ended up in Tucson in the first place, and he self-medicates to keep going. With parts of diseased lungs removed, he has the choice of sitting in a chair or taking life-shortening drugs that enable him to live a heroic life. Like Franklin Roosevelt, he is basically an invalid, but nobody knows it. One day toward the long ending, I walked into the house in Tucson, and he calmly said, "I've got myeloma, and I don't like that."

Sometimes I slept in that back room—unpainted cement floor and walls, plain thick wood shelves stocked with supplies and canned goods, bags of beans and potatoes, blankets, Seri crafts and artifacts on the walls, medications, paintings by John Houser and some by Cathy Moser. Flashlight always by the bed.

I am up well before sunrise and as always go for a long walk along the shore before breakfast. The beach drift is piled high with a fantasia of seashells and marine life. Paper-thin bleached piddocks, sea urchin testa, and coralline algae, bubble shells, pectens, tiny perfect sand dollars, coffee-bean cowries—the whole list is longer than this chapter. There is no way to avoid crunching them, but their life is gone. (The beach in Southern California where I grew up was also once like this.) At this time of year lots of eelgrass and winter seaweeds make up most of the beach drift, and I notice the first plastitrash.

I used to also go out for a long walk right after breakfast, when Ed would read from the New Testament. One day he informed me that I was the only visitor who did not stay for the Bible readings. So I put off my walks until after his reading. There was a clarity and calmness to his readings that I later learned were a mark of deep spirituality. Ike was truly interested and would engage Ed in Bible discussion, although he would admit to no organized religion.

Ed and Becky of course were passionate about the New Testament—their purpose for being at Desemboque was to translate the Bible into Seri. Along the way they published scholarly works on Seri culture—often but not always coauthored with specialists—made Seri a written language, taught Seris to read and write in their own language as well as in Spanish. They were forever on call to take someone in need to the hospital in Hermosillo, fix the water pump, and so forth. There was never-ending action, and I don't know how Ed and Becky put up with all the commotion or got so much accomplished. Everyone going through stopped in on them, including of course us.

The Mosers were the most tolerant people I ever knew. Over the years we maintained a close friendship although I sharply disagreed with many of their political views. My life has been vastly enriched by Becky and Ed.

My life has also been vastly enriched by Ike and Jean. They influenced my whole career. Ike and Jean contributed so much to my work, and to the work of just about everybody else I knew doing research in Sonora and sometimes Baja California and even South America, Africa, and Madagascar.

Ike knows the desert. He knows the sea. He knows the Sierra Madre. He knows ranching. He has vast practical knowledge of the Sonoran Desert and people and places. He can fix anything. But you better be prepared for Ike Russell time and you better like plain Ry-Krisp crackers, because that's often all we have to eat.

But I learned to be wary of landing too late in the day, too close to darkness or even in the dark, and try to devise ways to avoid such events. There was the time we were coming into Bahía de Kino a bit late in the day. We are just going from Desemboque to Kino to deliver some pet guinea pigs Jean is sending to Doña Gavina. We are coming in for a landing in a strong wind when I spot utility poles and wires strung all over the place. They weren't there last time. Ike pulls up fast and we search for another landing place. The wind really picks up and dust nearly obliterates the view. Landing into the wind means landing into low-angle blinding sunlight, albeit through murky dust. We can't be sure where the wires are and keep searching for a place to land to no avail in increasingly poor visibility as darkness descends. Now it's too dark so Ike keeps circling—a message to meet the plane. Finally someone savvy to the ways of small planes drives out to the paved highway and shines headlights in the right direction. We land and taxi in front of the headlights the half mile or so to Doña Gavina's little home and restaurant by the beach. As we get ready to leave in the morning, a cop drives up and hassles Ike for improper landing behavior.

Another time we flew out of Kino after a week-long trip to Isla San Esteban in the *Ofelia*. It's the old story—it is a bit late but we can always stop in Desemboque or find somewhere to camp. Now it's getting dark, so we land on a convenient dry lake bed. Oh well, a nice campfire and plenty of food. But the place is barren and not a scrap of firewood in a cold biting wind. We didn't think to tell each other, but I had left what food I had with Doña Gavina; Jean had done the same and so had Ike. To make matters worse, our sleeping bags were way too thin. While we shiver, Ike says, "This qualifies as an emergency." He finds a small Phillips screwdriver in one of his metal ammo boxes and methodically starts taking apart the backs of the front seats. Tough little plastic lab bottles full of tequila are securely duct-taped end to end to the tubular framework.

It's time to leave Desemboque and head for Altata. Becky and her guests drive us out to the airfield over the sandy road. In the usual bright clear sky we finally take off for Altata. You might wonder why I use the word "finally" so many times. When you travel with Ike it is on his time. He is methodical.

He is not like the rest of us. Nothing makes him hurry. You accept his pace or you better stay home, which is no fun compared to going on a trip with Ike.

We fly down the coast south of Guaymas. Hours and hours of wide green ribbons of mangroves clog long lagoons. I don't see the anticipated floating eelgrass rafts. In northern Sinaloa there are long rows of big beach houses on sandy shores. Like a Fellini movie, most of them are silently falling into ruin. Maybe it was a hurricane. The scene keeps repeating between more mangroves, and then we land at Altata. It is near dusk as we set down on coastal salt flat dampened bare ground in a wide place in the road at the edge of town. We climb out and stretch. Coconut palms lean over dilapidated houses strung along the strand above a broad waveless beach on an inlet bay. There is no breeze and everything smells moist and tropical. Nobody greets us. Tom and I walk down to the beach and look for eelgrass in the drift but only find *Ruppia*, another kind of seagrass. A nice botanical record but not what we are looking for.

Ike stops at a few houses and asks about a place to stay and buy meals. Nobody offers information. That's strange—even in the smallest place in Sonora there is always an older woman, often a widow in black, who fixes meals and lets travelers use an extra room or the porch for a modest donation. It's getting dark.

I walk over to some fishermen hanging out next to beached pangas and ask about eelgrass *zacate del mar*, describing it in detail. They act like they don't know what I am talking about. They're not like fishermen in Sonora. They don't laugh, not even at me. This place is creepy. Too late to go anywhere else. A bunch of gun-toting teenage soldiers announce they are guarding the airplane, which means we will have to pay them.

Finally, a gringo I will call Joe materializes and offers us meals and a place to stay. He lives here with no visible means of support. Like the rest of the house, the unpainted wooden floor sags, and the view is obscured by grime-clogged rusting screens. The old woman who serves us doesn't speak. I go out to the beach again and talk to fishermen still hanging around their pangas. We talk about sea turtles and again I ask about *zacate del mar*. They remain reserved and only slightly more friendly than before.

It has been a long day and I am thankful for a safe place to put a sleeping bag. The place is unbelievably dark and quiet. It seems like a different country from Sonora. In the first light I search the beach drift and skirt a mangrove lagoon, but there is only *Ruppia*. While I am at the beach, Kim and Ike talk to

Joe. I don't want to stay here. There is no eelgrass, and we would make better use of our time going back to Desemboque.

Joe didn't know jack about eelgrass last night and now says he knows where it is and can take us there in his pickup. I think he is lying, but what for? I get out our maps and ask him to show me where the eelgrass is. He looks at the map of the Altata region for a long time. Maybe he doesn't know how to read maps. He's a strange duck but then I suppose we are too. Why would anyone come here looking for something called eelgrass? Joe shows us a place way back in an *estero* (lagoon) where he says eelgrass grows. I know eelgrass does not grow in stagnant backwater. The place he points to consists mostly of barren dry salt flats we saw flying in yesterday. I ask how long it takes to get there. An hour or so. Well, I am sure that means at least two hours, and there goes the whole day for nothing.

I tell Ike and Kim I don't want to waste the time and expense. Tom does not say much. Both Ike and Kim are on my case. "What makes you think you know so much? This guy lives here and knows the place. You don't know everything." Something has happened and I won't go. Since I am paying for this trip I insist we go back to Desemboque. Ike and Kim put up quite an argument for going and seeing what Joe has to show us. This is one time I won't give in. I know about eelgrass and it can't grow there. I prevail and we put our things together. Still rather early and time to get to Desemboque at a reasonable hour.

Joe is really trying to persuade us. What the hell does it matter to him if we find eelgrass or not? Why does he care about it now? Oh well, maybe he is just lonely for American company or wants to earn a bit of money. We pick up our gear and head for the plane. Just past a muddy little alley, one of the fishermen I was talking to last night darts out from behind a house and says, "You guys better get out of here as fast as you can" and runs off before I can say anything. I catch up with Ike—he's taller and always walks faster.

I tell Ike, and Kim says, "What for? Didn't you ask him why?" I repeat we should get going, but Ike won't rush.

Ike checks the oil and fiddles with things under the cowl. The plane sits on a wide place in the road, next to a schoolyard full of kids. The soldiers guarding the plane have evaporated. "Well, at least we don't have to pay them," says Ike. Tom, Kim, and I climb in. Kim has a pilot's license, so he gets to sit in front with Ike. Out of nowhere a single-engine plane swoops down—pow . . . pow . . . pooff . . . pooff . . . pooff—the place is sprayed with bullets. I yell to Ike, we need to get the hell out of here—that fisherman knew something for sure. Ike closes the cowl, climbs in, starts up the plane, sticks his head out, and yells the usual "Clear!" He radios to the other plane making another pass at us. No answer. More strafing. Dirt blows up in little puffs at the same time we hear the "pooff . . . pooff."

The schoolyard is empty. Ike taxies out, keeps trying to radio, and we are off fast and short. Now I remember Ike and Joe talking about the plane last night.

Whew . . . we've left Altata none too soon. From my cramped seat, I look down at endless coastal plain of thornscrub and mangroves fringing the sea and look up to see the other plane coming at us. There he is right next to us. I am almost fascinated as the window gradually opens upward and in dreamy slow motion a steel cylinder slips out and spits fire. Shit! I am looking into a gun barrel firing at my eyeballs. I yell, "He's shooting at us!"

"What makes you think he's shooting at us?" says Kim in his ex-heavyweight-boxer studied macho manner. Bullets whiz in front of the windshield and Kim becomes a believer. Ike calmly says, "I don't know what I can do about his shooting at us, but I know what I *can* do." He plunges the plane, red-lining the tach to just above the water, wheels almost in the sea, and slowly outdistances our adversary.

We keep going north, stop for fuel in Guaymas, and head home. We find out Altata is the drug center for the whole west coast.

Ten years later: Sandy Lanham, who has a flying service out of Tucson for environmental scientists, takes a Mexican biologist working on Brant—the little goose that feeds on eelgrass—to the Sinaloa coast near Altata. He collects a big bunch of eelgrass and has Sandy bring it to me to verify.

9

# LEMURS AND ELEPHANT BIRDS

IT WAS 1968. Paul Martin asked if I wanted to come along on a trip to Madagascar. Of course. There was the matter of my job as senior curator at the Los Angeles County Museum of Natural History, but it did not take much to convince the administration I would bring back worthwhile collections and do worthwhile research. We would go to the arid desert-edge south end, a place I thought analogous to Sinaloa and southern Sonora in Mexico, the dry end of tropical deciduous forest, the thorn forest. Transition from tropics to desert was a strong focus of my work.

A quick look at species lists and I saw friends, succulent species of *Aloe*, *Alluaudia*, *Euphorbia*, *Kalanchoe*, and *Pachypodium*. But many of the species were not little plants in pots but full-grown trees the likes of which I had only seen in books. I got immunizations, film, a French–English dictionary, antimalaria pills, and a disguised money belt. Before leaving, old Herbert Friedman, the museum director who prided himself on being an old Africa hand (I thought colonial and tired), called me to his office and handed me a scroll tied in red ribbon. What is it? It's a "dazzler." A what? A dazzler. What's it for? Well, if you get into a sticky situation it might come in handy. It had legal-diplomatic speak in elaborate large print on impressively thick paper of golden glow faux parchment embossed with County of Los Angeles seals, some Latin phrases, and a bunch of ribbons like you see stuck on old soldiers' chests. I thanked him and never intended to use such an absurd thing.

Pierre, a recent PhD, Paul, and me. Pierre and I would meet Paul in Madagascar. I arranged for the two of us to stop in France to visit the Jardin Botanique Les Cèdres at Saint-Jean-Cap-Ferrat, Monsieur Julien Marnier-Lapostolle's private botanic garden. He was a gracious host. Cap Ferrat, a small peninsula on the Côte d'Azur, is nearly surrounded by water and almost frost-free. The enormous château, the most expensive home in the world, had impeccably tended gardens cascading down innumerable terraces to the blue Mediterranean, where a white yacht moored at the private dock.

There were twenty-five heated greenhouses, twenty-five gardeners and specialists, and 14,000 species under cultivation. Included were a warm orchid house, an intermediate orchid house, a cool orchid house (for highland tropical species like *Dracula*, *Masdevallia*, and *Odontoglossum*), a rainforest bromeliad house, a dry/tropics bromeliad house, several cactus houses, and the main object of our visit, the great Madagascar collection housed in a very long greenhouse. A botanic garden that exceeded expectations. Marnier-Lapostolle was interested in obtaining information and additional sources of plants from Madagascar. He was the largest liquor distributor in Paris.

Curt Backeberg worked in the Library at Les Cèdres, describing new cactus species often based on not much more than enthusiastic intuition, much to the chagrin of card-carrying taxonomists. Backeberg knew nearly every species of cactus on earth, and his unconventional taxonomy was sometimes of merit. His six-volume, 4,000-page *Die Cactaceae* (1958–1962) is a classic. When I checked out all six volumes from the University of Arizona Library, the librarian asked if I was going to read all that German. No, not really, just the Latin.

We went on to Athens, made like tourists, and the last evening in the plaza in front of our hotel the lights went out, steel shutters slammed shut on shops and cafés, and everyone disappeared. Gunshots rang out. My first insurgency. The next day we headed for Madagascar on Air France with a bunch of stops. Paul met us at the airport. The plane seemed bigger than the airport.

Antananarivo, the Malagasy capital, runs up and down hills of red rusted tin roofs and low French colonial buildings and an occasional small high-rise. Streets and roofs fresh washed in a late afternoon tropical rain of soft sunlight. Papayas and mangoes and straight-backed women carrying babies and baskets and chickens and Malagasy music and speech among the French. I wandered through the marketplace in the narrow streets below the hotel and lost count after more than two dozen kinds of bananas of diverse sizes and colors. Antananarivo, a magical postcolonial miniature. The new republic still tightly gripped by France.

Food was French with enough Malagasy spices and tropical fruits to make it seem like heaven, and good wine was cheap. Next day was the herbarium, botanic garden, and zoological garden and researchers therein. The herbarium had a French inheritance of impracticality. Specimens tied up in bundles and in open-air wire bins that went way up so you needed a ladder. There wasn't enough light, and data on the labels were often scant and not useful. Yet I got to see distinguishing features of key species. Many arid-land species were not in the botanic garden because it was too wet. Paul had long conversations with some of the zoologists, and it was fun to see so many lemurs.

The four-hundred-mile flight southwest to Toliara in a prop-engine Air Madagascar (the name recently changed from Mad Air) took us over eroded red land of no towns. Two thousand years ago early immigrants from Asia burned the forests and the topsoil slid into the sea. Toliara (formerly Tuléar) is the only town on the southwest coast. We were transported in a Land Rover to the French research station at the edge of town. The place was nondescript and basic except the two-hour French lunches. One time a ceiling gecko missed its fly and fell into the soup tureen.

Not far from the station, a low hill of whitish gravel bore low clumps of several succulent *Euphorbia* species, some named by Backeberg. My first encounter with *Euphorbia stenoclada*, one of the many leafless euphorbia trees, but this one has sharp spines that no matter how careful you are, rips your clothes and sometimes flesh. The gravel was a midden heap of broken elephant bird (*Aepyornis*) eggshells. The first immigrants feasted on these huge eggs.

The American Museum of Natural History wanted an *Aepyornis* egg, the largest single cell ever, from the world's largest bird—a ratite, like an ostrich, weighing up to one thousand pounds and said to have inspired the biblical legend of the roc. The locals used the eggs as water bottles as late as the nineteenth century. These were from *Aepyornis maximus*, the elephant bird, the largest of the perhaps dozen or so species known only from Madagascar and extinct soon after people arrived. Every so often shifting dunes exposed a whole egg. One or two were available for around $800, but the American Museum would only pay $500.

We drove across the arid south end of this miniature continent on dusty roads, narrow tracks across arid plains, sometimes all day and not seeing another person, let alone a vehicle. The villages had no electricity. We stayed at so-called hotels, less than basic, and were the only guests. Narrow cots in one large room upstairs, plain unpainted weathered wood, windows without glass. Oilcloth covered a rickety table downstairs. Dinner was rice, sweet termites, and an occasional small chunk of meat. Sometimes we advanced funds for the proprietor to purchase the rice. Dark after sunset, so not much else except to try to go to sleep.

One full moon of insomnia, I heard drums. I put on my shoes, crept downstairs, grabbed a big dog stick, and headed out. You carry a big dog stick because the big dogs are not friendly. I wandered into a clearing of drums and a gigantic bonfire. People in a circle on small wooden stools as endless tales were recited. Stories of the honored man of the funeral celebration. I wandered away as dawn approached and sat wide-eyed on a sandy bank, against a thick-trunked tamarind tree. A troupe of twenty or so *Lemur catta* ambled single file down an adjacent limb and walked in front of me in no particular hurry. These ring-tailed lemurs held long tails flag up, banded black and white raccoon-style. Long snouts like a coatimundi. Every so often one turned to play or swat a teammate. A few minutes later a troupe of ten sifakas came along the same route, their routine lively and dignified. The largest living lemurs, they walk upright. A long white tail, white body, and humanoid faces with big eyes in a black bandit mask. They tumbled and jumped, and one leaped from above to land almost at my feet. I remained motionless as several of them gave me a close inspection and then bounded off to morning activities. I wandered back and climbed in bed just as Paul admonished me to "get up, lazybones."

Red dust enveloped the Land Rover as we drove southeast from Toliara across the Betioky-Ampanihy plains of low scrubby thornscrub of spiky euphorbia bushes. A few meager acacias and colonies of a *Stapelia*, a member of the milkweed family that produces starfish-like fly-pollinated red-brown fleshy flowers that stink like rotten flesh.

A young Mahafaly guy came out of the bushes with his big metal spear and wearing the usual brief loincloth and a smile. Unable to exchange a word, I offered cigarettes, which made him smile even more. The government has draconian laws against growing tobacco, so it can get taxes. Locals crave cigarettes they cannot afford. I wish I could have talked to him about plants. I gave him the rest of the pack.

We see a tower on the horizon. Turns out to be a giant baobab. This tree has a straight thick bole unbranched for about eighty-five feet. Thick stubby branches stick out on top of that gray-brown shaft. Leafless now in the dry season. At ground level on the north side is a hole, barely big enough to wiggle through. Once a door, the softwood trunk has grown around the entrance, making it ever smaller. Inside the trunk is hollow like a cathedral all the way up to the branches. A hole near the top lets in a shaft of dusty light. Disk-shaped stones, each about eighteen inches across, sit neatly and evenly against the wall all around the room. Paul thinks he might find bones of extinct animals in the floor, animals alive when the tree was young. Hanging from the ceiling is a shit-bombing flock of endemic mega-bat flying foxes. Paul's mission to the island was to find evidence of human-caused extinction, to support his theory of Pleistocene overkill. Madagascar would be one of the best places, since people arrived late, and megafauna extinctions promptly followed.

Paul is ever asking after fossils, old bones. At one Mahafaly village, he asked a bright-looking young man if there were any old fossils around the village. "Yes, and do you want me to take you there?" Narrow alleys between high mud walls, and we seemed to be backtracking. Eventually he led us to a hut. Pushing aside the coarse woven curtain of a door, he took Paul inside and introduced a withered old woman who gave Paul a wide toothless grin.

At a small village we were hauled into the police station. One not large room, walls once white, window bars and no glass. Jail made up half of the room. Plastic chairs and a gray metal desk with no papers. Overweight Monsieur Police Chief wore a blue uniform shirt and a cap befitting his station. I tried not to be looking under the desk at his bare feet on the bare cement floor. He sat; we stood. He questioned and was not satisfied with answers. Maybe he was put off by Pierre's Parisian French. We were in trouble. I had my field pack and handed him the "dazzler" old Herbert Friedman had given me at the Los Angeles Natural History Museum—"if you get into a sticky situation it might come in handy." Monsieur Police Chief held it for a long time, upside down, pretending to read, and finally handed it back, shook our hands, and wished us well in our travels.

Zebu cattle are Mahafaly wealth. Men's lives are spent accumulating cattle for their funeral sacrifices to feed the guests. The richer the man, the more cattle, the longer the funeral. We met a man with cancer who had sixty-four wives; the whole village was his. Already they were talking about his funeral being the biggest in the west, going more than six months. He had a dozen or so watches on his arms and rode around inspecting his village on a motor scooter.

On our way back to Toliara, the vegetation became lower and more desert-like, but I would not call it desert, perhaps closer to thorn forest. To the north it was like the dry tropical deciduous forest in southern Sonora of northwestern Mexico, but a change of species and more succulents. Madagascar does not have a thorn forest like the dry tropics of adjacent Africa. Most trees and shrubs have much shorter spines, rather more like a prick forest because the original animals were smaller and smaller spines would be good enough.

There were some notable exceptions to the small-spine rule. *Pachypodium geayi* is a succulent pole with large spines in the same family as oleander, with a head of leaves present only during the hot summer rainy season. At other seasons this thick pole is leafless. I saw *P. geayi* sticking up through the dense prick forest of dwarf tree aloes, ocotillo-like didiereas, strange euphorbia trees, fat-bottom adenias, and occasional pseudo-bulb ground orchids resembling orchids that usually grow on trees in the tropics—botanical wonders I knew only from potted plants and photos. In my enthusiasm I got clawed by a *Euphorbia stenoclada*, and pulling away I crashed into a bunch of *P. geayi*. For many painful hours I realized they have toxic spines to protect their succulent stems.

One day a dusty track through the Betioky-Ampanihy plain brought us to Ejeda, a village with a Lutheran missionary hospital. Dr. Curt Stolee and his wife, Peg, originally from Minnesota, ran the clinic and small hospital. They invited us for lunch complete with canned peas and then asked if we wanted homemade ice cream. I realized what deprivation might mean to an addict. Like other missionary people I have met around the world, they were a lot of fun and hospitable, and offered fascinating local information. We said goodbye and I thought about the strangeness of saying goodbye to new friends never to be seen again.

I made vegetation quadrats to document species diversity, the numbers of species in an area, and the growth forms making up the vegetation (number of tree species, shrubs, vines, etc.). I found the numbers consistent with those of the Sonoran Desert as one goes from tropical deciduous forest into arid places, that the number of species toward aridity declines in a predictable manner. The quadrats were twenty meters by ten meters, each plant counted or estimated. Some ecologists spend their lives doing quadrats.

We returned to the French research station in Toliara. I was contemplating new places, such as Andre Diome's lemur reserve, the lush tropics of southeastern Madagascar, and the rocky highlands so rich in strange endemic plants.

Among the botanical attractions of the Toliara coast is the rich assemblage of seagrass species—seed plants growing on the shallow sea floor. We went out in a rowboat. With a face mask I saw half a dozen species of seagrasses. The sea was warm and crystal clear, with bright green seagrasses on white coral sand. Pierre was picking up corals, which I did not approve of. Underwater he handed one to me, not seeing a long-spined sea urchin on the backside. The spines are long, thin, sharp, and brittle, and they penetrated right through my hand. Painful as it was, the spines dissolved. I went on diving, holding my breath, discovering more seagrass. I have always been a good swimmer; I grew up by the ocean. But I was out of practice. Suddenly I felt an unmistakable stabbing pain. Unmistakable because I'd had the same thing a year earlier during exertion at high elevation in Colorado.

Back at the research station I told Paul I had a collapsed lung and needed immediate medical attention. After even a short sentence I was gasping for air. Paul told me to calm down, I was probably just out of breath from diving. I explained that I knew what a collapsed lung was like and it could be life threatening. I asked Paul to take me to the local hospital, but he and Pierre said not to worry, they would take me after lunch. Every breath difficult. You might get by on a collapsed lung for days, and sometimes it heals itself, or it could be hours or minutes until you die. I again told Paul to get me to the hospital.

After lunch they were about to drive me to town, but Pierre said he was hot and sticky and needed to take a shower. Finally, we drove into town. It is a simple operation—insert a tube into the thoracic cavity, pump out the air that has leaked out of the lung, and the lung reinflates. Depending how long it takes the lung to repair itself, air is kept pumped out of the thoracic cavity. The modern multistory hospital was staffed by Malagasy doctors. The surgeon was on duty but would not do the surgery. Not on a white man. What if something went wrong? No, I had to go elsewhere.

The only thing I could think of was to hire a plane to the missionary hospital in Ejeda. I asked Paul to take me to the airport and find a pilot, but instead he went back to the field station and said the people there would make phone calls to find a pilot. The pilot was out and would call back. I was experiencing difficulty. I remember staggering, careful not to fall, into the station where I found Paul and told him I might make it to the next day, but I would die if the pressure collapsed the vena cava too. The afternoon was wearing on. This time it got through to him or maybe it was luck, but a pilot was located, and Paul and I were on our way to Ejeda. I repeatedly asked the pilot to fly lower; we were cruising too high for my condition.

Curt Stolle met us at the little landing field by the hospital and in as few words as possible I told him what happened. He wasted no time. Later he said he was concerned because my lips were blue. He had never performed such an operation. The last thing I remember was seeing a red fountain and then I woke up with plastic tubes sticking out of me and happy to be breathing easy. Paul returned with the pilot. I was regaining strength and would go home soon. But my hosts had to leave Madagascar for a few weeks and said they could not leave me there. The plan was to go by air ambulance to the French military hospital in the capital.

The last thing I wanted was another plane ride and to be stuck in a hospital. But I was in no shape to protest. Curt gave me morphine and they put me on the floor of the plane. By the time I got to the hospital in Antananarivo I was barely conscious, with a thoracic cavity filled with fluid. Holes were poked through my back, fluid drained, and my lung was functioning. During the ensuing days, I had reactions to large doses of intravenous antibiotics. Spirits and strength draining. Paul paid me a last visit, said "you'll be okay here," and took off for East Africa. About a week later I thought I was recovering. A young French doctor took me out for a real lunch, but the rich food and overdoing set me back. I contacted the American embassy, and all they did was make sure I was able to pay my bill. An American minister came by with a Bible and said Christ is the answer.

One night a barefoot nurse gave me a handful of pills I had not seen before. I pretended to take the pills and got rid of them. I woke to someone going through my belongings and thought to call for help but the help was doing the pilfering. French doctors left at night. I kept quiet. I had left a $20 decoy in a pants pocket. I had more money in a belt that did not look like a money belt.

Finally I persuaded hospital officials to let me go and got a few morphine suppositories they call pink rockets. They take everything by suppository. A taxi to the airport nearing 11 p.m. on a rough bumpy dirt road and I was leaking from the last incision and had stomach cramps from antibiotic toxicity. If I looked sick they would not let me on the plane. I managed a pink rocket in the taxi. Airport officials kept me for last, and the plane was about to leave when at the last stop before boarding, customs official says, "Your visa has expired. You can renew it downtown tomorrow." When did it expire? Midnight, and it's after midnight (12:20 a.m.). But I am leaving (and the visa was not expired). Sorry. Regulations. *Combien? Tout ce vous voulez, Monsieur.* All I had was a $100 bill, and I never looked back.

In Paris I was supposed to go to the American hospital. No more hospital, no matter how clean and wonderful. "Only one seat available." First-class bargain standby nonstop Air France to Los Angeles leaves in half an hour. Americans in the seats in front of me complained about the food. I was on my way to recovery.

# 10

# WILD LIFE OF KENYA

OUR TENTS ARE PITCHED between a wall of Senegal date palms and fever-tree acacias. These aren't clean, trimmed city palms. These are wild Senegal date palms (*Phoenix reclinata*). Dead leaves and dry stalks decay beneath long shiny green leaves, and multiple reptilian trunks branch at ground level, curving to the Kenyan sky. All interwoven with dagger thorns of modified leaflets at the base of each palm leaf, along with tropical lizards, spiders, and squawking birds.

All night a pride of lions belch and slurp guts and crunch bones from their wildebeest on the other side of the palms. It's the middle of the night and I see ghost columns moving around while things drop onto the tent. Giraffes are stepping around and over the tent ropes, chewing fever-tree pods and twigs, dropping thornflakes. How can they see the slender ropes so far below and what if one trips? The lions keep up their snarling, and I forget about going out for a piss.

At dawn the lions are gone, and we are on little stools drinking tea looking across a dusty vista between galleries of fever trees. We are privileged scientists at the research camp. Jeashi, the Masai ecologist who graciously introduced us to this scene, politely suggests moving closer to the tents. We are finishing breakfast when I notice dust on the horizon and a herd of elephants bearing straight at us. They go right through where we had been sitting. The ground shakes. Jeashi says they come through here every day at this time. Elephants are the closest you can get to the Pleistocene and you never forget elephant encounters. Rhinos might qualify but don't seem sentient.

In the early morning air Kilimanjaro looms snow clad over the southern horizon in neighboring Tanzania. Dry-season air soon hazes the near horizon and the elongated white peak floats over the savanna.

I change to shorts and sandals, realizing why everyone else does the same. It's so you can see the ticks walking up your legs and get them off before they get you. At the game parks they don't let tourists out of the *kombis*, not because they might get eaten by lions but because of ticks. With all these big mammals the grasses are full of them. Many times I see a flat seed tick, a tiny dark disk, sitting on top of a dry grass stalk. As I move closer, it starts waving little claw feet to grab any mammal hair that might brush by.

This is East Africa and lawyers don't go to court if your loved one gets eaten by lions. We are told not to wander around at night, and best not to wander around alone even in daytime. A few weeks ago a researcher was walking alone from the lodge to the research camp after dinner. She was never seen again. There were plenty of lion tracks.

We rented an aging Volkswagen bus in Nairobi and it is cranky about starting. We are on our way to one of the salt pans in search of elusive halophytes of economic potential (my grant funding) and there is a big rhino in the distance. As I fiddle with the telephoto the rhino comes closer. It's a boy and he heads right for us. He has a huge horn and is bigger than the Volkswagen, and it won't start. Engine in the back so you sit right up front. The rhino ambles up to us, turns around, and rubs his butt on the VW, rocking us to and fro, then lifts his tiny tail, drenches us in rhino pee, and saunters off. The park people know him by name and say this is his usual manner.

We are back in Nairobi working at the National Herbarium in the Natural History Museum of Leakey family fame. This is our headquarters for identifying specimens collected during the last few weeks. We moved to a pleasant old inn within walking distance of the museum. A welcome change from the first night in Nairobi.

In Tucson, Paul Martin and Ike and Jean Russell recommended a place in Nairobi. Before leaving Mumbai, we cabled to make reservations. It's dark when we get there and they say it's the only time they ever received a cable. We are the only ones in a drab dining room that doesn't look like a hotel dining room. We are hungry and there is no menu. They ask how many kilos of meat we want. There isn't any other choice and we have no idea where any other eatery might be. We don't want a kilo, maybe one-eighth kilo. "Minimum order one kilo." It isn't expensive, and we really get a whole kilo of roasted meat, and it isn't bad. We leave most of it and I am sure it will not go to waste.

But the exposure to ether on the flight must have done something, and I spend hours in close communication with the toilet, which mercifully works. Yes, ether. An unsavory looking couple in the seats in front of us had liquid ether. People were nodding off, and fear gripped me as I leaped up front to the purser and told him to warn the pilot that there was ether on the plane and to make people stop smoking. Yes, across the Indian Ocean people were smoking. Some Brit said, "calm down, old boy." I told him to shove it and demanded the purser find the ether and secure it in the freezer. Took minutes to convince him and find the couple; they were dazed. She said it was her nail polish remover. A whole jar? More people were nodding off. Purser asked her to put it in the plastic bag he provided and asked if I was satisfied. NO. This is liquid ether and it can explode. Take it away, put it in the freezer, lock it up, and guard it. Do not open the freezer until landing and everyone is off and all plane doors open.

When finally into deep sleep at our mysterious out-of-the-way drab hotel, the place erupts in high-pitched female voices and loud rowdy men chasing them. Two a.m. Lots of giggling and unmistakable sounds from adjacent rooms. Next morning we learn the rooms are rented by young Italian *artistes* who work at the casino next door and entertain rich African men after hours. We find our way to the museum and quieter surroundings.

Polite young African men, in business suits and well educated, are asking for handouts. Ugandans fleeing Idi Amin, the dictator who says Hitler is his role model. There are no jobs in Nairobi. Also, legions of Kenyan young men, the third, fourth, fifth, sixth sons, who have to leave the family farm. Shrinking plots can no longer be divided. No clue what young women do to survive.

Nairobi's climate is sweet, an intermediate elevation in the tropics. No frost, pleasant by day and not hot. A horticultural paradise. I enjoy the walk from the hotel through the park to the museum. East African cycads (*Encephalartos hildebrandtii*), some reaching five meters, tree aloes in full flower, and giant strelitzias, the real bird of paradise plants. The museum staff is shocked that we walk through the park. "It didn't used to be like this but now we have so many desperate young men. You need go by taxi." A taxi for three blocks? After lunch a few days later, a botanist from Switzerland staggers in dripping blood. Mugged on the cycad park path.

11

# NYET IN THE USSR

## FOURTEENTH SESSION OF THE GENERAL ASSEMBLY OF THE INTERNATIONAL UNION FOR CONSERVATION OF NATURE (IUCN) AND FOURTEENTH IUCN TECHNICAL MEETING / SURVIVAL SERVICE COMMISSION, ASHKHABAD, USSR, SEPTEMBER 26–OCTOBER 5, 1978

I STEPPED OFF the plane in Moscow, and it was snowing. It was early evening, and I got shuttled to a short customs inspection line while the rest of our party went quickly through longer lines. I was behind a stout young woman going to Russia as an exchange librarian. Cold War academic exchanges were considered safe and controllable. She had a suitcase full of copies of her new book. Those unsmiling inspectors went through every book, page by page. I tried to change lanes. *Nyet*. They finished after midnight and did not even open my suitcase. Welcome to the Russian Soviet Federative Socialist Republic.

The librarian and I found a taxi to our hotel, the Rossyia on Red Square, the biggest hotel in the world. I was issued a special hotel pass card and told not to lose or forget it; without it, no admission, even if freezing cold. The world of *Nyet* was unfolding.

Next day was sightseeing with IUCN Survival Service Commission friends. Red Square. Babushkas stooped over sweeping prairies of cobblestone with short-handled brooms of bundled sticks. We walked along wide avenues eerily quiet with unbelievably sparse traffic. No smog. Occasional dark-windowed black ZiL limousines zoomed by. No siren, no stopping, everyone knows to get out of the way. Our little multinational, multiracial group must have stood out. A big car screeched to a stop in front of us. Two guys jumped out, ran up, and asked if we had any pot for sale. Of course not. The Keystone Cops run back to their car and zoomed off. I didn't think to say *Nyet*.

We gathered at cocktail time and were deciding about dinner when a tall wildlife expert from an African nation asked if we would wait for her; she would be right back. She had a bundle of new Levi's jeans, coveted status symbols. She was selling them to a man in Red Square. The papers were full of examples of Westerners arrested for contraband, and someone reminded her that her buyer could be an undercover agent. "Oh no, I asked him, and he said he wasn't a policeman." She was soon back with a thick wad of US$100 bills.

Up and packed at 4:00 for a 9:00 a.m. flight to Ashkhabad (Ashgabat), 1,500 miles southeast of Moscow on the Iranian border. Had to be four hours early for a flight. Sheremetyevo airport for internal travel. Flight delayed. When do we leave? Soon. Can you tell us when? *Nyet.*

The main waiting room was one gigantic square building filled with thousands of dreary people. Travel in the Empire was free, everyone supposed to get a vacation, but the ordinary folks wait for the next available flight after applying for travel permits. Not uncommon to spend half your vacation trying to get to where you were going. Looked like some folks had been there for days. No sign for the restrooms; didn't need one, stench unmistakable.

We were in the cozy VIP waiting room and had our own restrooms. We could stretch out on fake leather seating in our unstinky cove. No barrier or private door from the main waiting area, but no one unauthorized wandered in. Plenty of time to make friends, interesting people. By afternoon the only place selling food was closed. I had been forewarned and most of my suitcase was full of snacks for sharing: cashews, peanuts, dried fruit, jerky.

Again, when was the flight leaving? In a few hours. In four hours. Soon and so forth and *Nyet*. We spent the night. Much later we learned it had to do with copy machines airlifted from IUCN headquarters in Switzerland. The meetings are conducted in at least four languages and the endless translations had to be available quickly. Public access to copy machines? *Nyet*. It was a few years before faxes helped bring down the Empire. Inappropriate use could threaten the government.

The Russians really wanted this IUCN meeting. A big propaganda coup. It was supposed to be held in a Third World country. The Soviets finagled occupied Turkmenistan to qualify.

Nearly dark. After a little while, nobody stopped me when I went out for a walk to breathe fresh air. If there were any signs, they were in Russian, and I didn't think I had gone far when it started snowing and Russian soldiers yanked me aside. I showed my airplane ticket and someone grabbed my passport. "Amerikanski! Amerikanski!" Guns stacked like cornstalks as more snow fell. "Amerikanski!" They pulled out vodka bottles from snow-flecked great gray coats. I was getting cold. "Amerikanski, drrrink!" Not much choice, and I realized they were drunk around a bonfire singing to vodka, guarding the airport. More vodka and I wasn't cold. It was a blur when I stumbled into the terminal, not sure if I was even going in the right direction. The flight hadn't left.

The next day we and the copy machines boarded Aeroflot for Ashkhabad. Someone called it Aeroflop due to credible rumors of recent crashes. I foolishly looked in the seat pocket in front of me. The contents should not be described. Stella and I were more than friends and I looked forward to an interesting time. That evening I headed for her room on the other side of the elevator that divided each wing of the hotel. I hadn't noticed earlier, but sitting at a desk between the two wings was a babushka who stood up and stopped me with a loud *Nyet*. I'm just going to see a friend. *Nyet!* Maybe she didn't understand English, and Spanish didn't work, and I knew not to try German in post-WWII USSR. I am just going to see a friend. *Nyet!* Single men and women were assigned different wings.

Not long after I returned reluctantly alone to my room, Stella knocked on the door. "How did you get past Brunhilde at her desk?" "We women have our way." Stella always had a twinkle in her eye. Ashkhabad is the transliteration from Russian, originally from the Persian, meaning "City of Love."

Workmen were still pounding nails when we arrived at the multistory hotel, built just for our meeting. The ubiquitous unsmiling Lenin statue had just been installed. The nearby meeting hall was likewise being finished as the sessions started. Fascinating and significant reports and information after getting through the endless speeches by Soviet politicos followed by IUCN bureaucrats. Stella discovered the best thing going in Ashkhabad. Along the London plane-tree-shaded street from the hotel to the meeting hall was a sidewalk kiosk where a smiling woman sold homemade ice cream.

The globally critical Survival Service Commission creeps along. The French: "Mr. Chairman, it is a matter of verb." The English: "Mr. Chairman, there is a split infinitive." The simultaneous translators say, "the infinity is split." While species hover at extinction, they bicker over rules and regulations far from the struggles of people displacing the planet's biodiversity. What to do for the last twenty-seven Sumatran rhinos? The English gasp and snort at suggestions of cutting off the horns that would help wreck their economic value. Desperation to stop the trading of turtles and crocodiles, monkeys and

The Fourteenth IUCN conference in session, Fall 1978. Richard Felger archive.

elephants; see how we carve up the planet. The daytime food is vile, so we eat ice cream.

The brand-new hotel was designed in Moscow and as usual the architects had not seen the site. Across the street was a substantial area surrounded by very high walls. Someone said it was a gulag. All rooms on that side of the hotel were blocked off.

At the end of the day our hosts treated us to sumptuous banquets, but you had to sit through endless speeches extolling the glories of Soviet achievement. They get people to be on time by putting out the finest wines and vodka and hors d'oeuvres, but never enough. I remember spending time by the caviar with Lady Scott. Poor Sir Peter, chairman of the Survival Service Commission, had to sit up front and stay awake without fine wine and caviar.

The North Korean delegates were thugs posing as scientists. They had recently killed South Korean scientists at a conference in Prague. The Russians were doing everything possible to prevent a repeat. Every night the South Korean delegates would be smuggled into Americans' rooms and the doors bolted. I made sure the in-group knew I wasn't the only one in my room.

America has lots of subcultures; Russia has listeners. Monolithic Soviet apartment blocks have an entire floor of listeners. If you look under the table in the few upscale restaurants, you see a bunch of wires. It's hard to talk because of the blaring music, so people can say private things. Smoke detectors in our rooms had multiple wires. In the privacy of their room, Lady Scott said to Sir Peter something to the effect that "if another Russian shoves another glass of vodka in my face, I will kick him in the balls." Next day no Russian tried to make Lady Scott drink vodka.

Bird watchers versus the KGB: The birders planned early morning outings for unprecedented life-list opportunities. The best nearest place was south of town, between Ashkhabad and the treacherous Iranian border. Off limits. *Nyet*. Being denied bird watching was not acceptable for well-connected birders. Cables to royalty and heads of government flew around the world. Sir Peter to Prince Philip to the Kremlin. KGB relented, "but on condition we go with you for your protection. We meet you tomorrow at 9:00 a.m.; that's one hour before the meetings start." No way. The birds are active at dawn; we leave at 5:30 a.m. *Nyet*. 9:00 a.m. Not acceptable, we go at dawn. And that did it; no hungover Russian security agent is going to get up for 5:30 a.m. So "all right, go look at your birds."

One evening a bunch of us went to the opera, which was well supported by the State, and even in this backwater capital, the singers were professional stars, although the sets and theater were shabby. We were almost the only ones in the audience, admission a pittance, but classical opera was not popular even if mandated by the State. (Contraband Western rock 'n' roll cassettes were ubiquitous.) Tonight's opera: "Boy Meets Girl by the Karakum Canal Driving the Collective Farm's Soviet Tractor."

Everyone was a specialist, well, almost everyone except the North Koreans and one jock-looking American. There was Ms. Freshwater Otter, Mr. Stork Expert, Ms. and Mr. Elephant Experts, Mr. Crocodilian Expert, and so forth, as well as well-known NGO bigwigs and ecologists. But this American dude was no expert. I asked where he worked and what was his interest, and he mumbled something about environmental policy at some Middle American college nobody had ever heard of. Our Russian hosts would pick us up for field trips and banquets and leave Mr. CIA standing blank faced in the hotel lobby. Once after returning from the botanic garden, Mr. CIA cornered me and wanted to know, "What kind of equipment do they have? What are they doing there?" and other clueless questions.

Mikhail was assigned to a few of us Americans, likable and helpful. KGB. He knew where I went to high school and had my bio down pretty well. After the usual amount of vodka at an Ashkhabad nightclub that seemed like a nightclub just about anywhere, he said the stupidest thing he ever did was to come back after he did his service in India.

Someone offered $300 for my Levi's jacket, "No, not rubles, dollars," opening his wallet to show me C-notes. The Iron Curtain was no barrier to the well-oiled well-known black market that ran on American dollars.

One afternoon was a field trip south of town into the mountains near the Iranian border. Ashkhabad sits on the southern margin of the great Karakum Sand Desert, and the southern border is a series of intriguing mountains. Just like the desert-edge grasslands of Arizona and New Mexico, the trees huddle along canyon-bottom riparian places. The steep rocky slopes were heartwarmingly brown with perennials, grasses, and some old friends like a stubby species of ephedra and a dwarfed pistachio shrub, as well as the rare Afghan fig, *Ficus afghanistanica*. At the botanic garden I was given cuttings of this rare fig and Stella smuggled them out of Russia under her blouse. I got the cuttings home in time and they took root and developed into handsome shrubs thriving in Tucson.

The best excursion was three hundred miles east to Repetec Sandy Desert Reserve, outside Chardzhou near the Uzbek border, now known as the Repetek Biosphere State Reserve, in the East Karakum Desert. As a result of our visit and the IUCN meeting, it became an International Biosphere Reserve that same year. It was a short flight from Ashkhabad to Chardzhou in the Lebap district. Walking down the street to our restaurant, people stared at us as if they had never seen Westerners, and of course we were more than conspicuous—Europeans, Latin Americans, Africans, Asians.

There were the usual faceless multistory slabs of Soviet apartment blocks, cement gray and balconies strung with laundry, but with some outstanding features. In this wind-blown sandy desert, there were no gardens or trees in front or back. Instead, in front of many doorsteps, fat furry brown two-hump camels sat sphinx-like waiting for something to happen. In back there were rickety jerry-rigged multistory wood and tin outhouses.

We were bused to the Sandy Desert Reserve—pale moving dunes like the Gran Desierto of the Sonoran Desert. The plants and animals were different, although ecological analogs: the northernmost cobra, the Central Asian cobra; the northernmost monitor lizard, called *zemzen*; agamas instead of chuckwallas; and the rare, endangered *Gazella subgutturosa*. We were taken to an area of high moving dunes in a straggling fleet of high multiwheel open military

IUCN conference participants disembarking from Soviet transport vehicles on their field trip to Repetec Sandy Desert Reserve in the East Karakum Desert of Turkmenistan, Fall 1978. Richard Felger archive.

Richard purchasing a slice of melon during the IUCN field trip to Turkmenistan, Fall 1978. Richard Felger archive.

trucks. Awaiting us was an array of mouth-watering melons, like you see in every Turkmen open market.

The dry desert made me feel at home. In addition to our military transport trucks were two elderly Soviet military ambulances with a big red cross painted on each side. There were a bunch of Russian women, apparently the organizers of the event, tottering around in the sand in high heels and low-cut city dresses. The Russian research staff looked like anywhere in the western United States, Israel, or Australia, knowledgeable and providing meaningful info. The 100-degree desert proved too much for the Norwegians, who turned pink and were treated by the medical staff.

I saw my first living *Ammodendron* trees (*A. conollyi*) cresting low dunes. Skinny legumes with silvery leaves reminded me of a cross between Sonoran Desert smoke tree and ironwood. But the outstanding elements were the saxaul trees (*Haloxylon aphyllum*). Fat succulent trunks, like Sonoran elephant trees, but leafless like an ephedra, the twigs photosynthetic gray-green. This endangered tree is a chenopod related to the invasive tumbleweed (*Salsola*) of western America. Stella and I sat in the sparse shade of a saxaul tree that capped a small dune and savored succulent slices of pale melon. A perfect afternoon on the other side of the world in a desert wilderness with Stella. One expects occasional desert mirages but this one was different. The ground in front of us was quivering and heading for us, a sea of long-legged ticks running toward our warm bodies.

One afternoon back in Ashkhabad, we were treated to horse races. Horsemen on the Steppes of Russia are famous, but the camel races were the real attraction. You wouldn't think those furry brown lumbering two-humpers could move so deftly. As the second race was underway, a wall of dark sky was moving our way. Time to leave. One day in five is a sandstorm like out of *Dune*. Traditional houses and stores have double windows about a foot apart to trap sand and dust—you see windowsills deep in sand. Before the time of Christ, the land was forested. Then horsemen torched the forest for grassland that was still holding at the time of Omar Khayyam. Later the grasses went with overgrazing, and even the microphytic crust gave way as dunes spread thousands of miles across the steppes. Russian researchers at the Repetec Reserve were having moderate success in stabilizing dunes with 1.5-meter waffle-like straw squares, allowing a microphytic crust to establish as microhabitat for new vegetation. But the effort seemed dwarfed by the magnitude of devegetation.

I thought it would be interesting to return to Moscow by train. *Nyet*. By bus? *Nyet*. Only by plane. Nobody allowed across the country except by plane. Military secrets. I forgot to tell you about the endless hills at the edge of the plains near the Iranian border mountains. Those hills were all the same size and shape, housing unseen weapons.

I hated it that Stella was always last minute making a plane connection. My Ashkhabad-to-Moscow ticket was supposed to be in her purse for safe keeping since I am always losing things, and she couldn't find it. Everyone else was on the plane ready to leave, and I had no ticket. Yikes, stranded in Ashkhabad. Solution: Run into the ticket counter and buy another ticket. "You can get your refund in Moscow when you find your original ticket." Sure, I could go stand in line somewhere for a couple of days to be told *Nyet*. Last day in Moscow she found the ticket in the bottom of her purse.

The Russians arranged for the South Korean delegates to sit up front among the Americans. As soon as the plane landed and before anyone else, meaning the North Korean thugs, could get off the plane, we got off in a tight group surrounding the South Koreans and were whisked into U.S. Embassy vans and sped off across the tarmac circumventing the airport terminal. The South Koreans were taken to the U.S. Embassy, and the rest of us to the Rossyia Hotel.

Stella had an introduction to a famous Russian nature filmmaker. He met us at the Rossyia in the late afternoon, and we were chauffeured across Moscow to the 540-meter-tall Ostankino TV tower, the central broadcasting nerve center housing the All-Union Radio Television Transmitting Station of the USSR Ministry of Communications. The nerve center of the Empire was well guarded. I lost count of the number of locked doors we passed through, each time clearing increased security with our filmmaker friend. In a small plush theater high up in the tower, we were treated to a private showing, highlighted by his film of polar bears swimming underwater close up without a telephoto lens. He had been a young boy during the nine-hundred-day siege of Leningrad. His parents had put him between them at night, and one morning he woke shivering between his parents, who had frozen to death.

After the viewing we were chauffeured across an endless sameness of Soviet apartment blocks. We halted at one that looked like all the others, went to an unmarked door and through multiple security checks, up an elevator, through another locked door and security check in a deserted maze, and down a faceless hall to double doors. Like stepping into a magic world we were in the Soviet filmmaker's club, a great open space of high-ceiling opulence the likes of which one sees in Beverly Hills, like Hillcrest Country Club, where the liberated Hollywood film folks set up a club of their own after being excluded from the restricted country clubs. The most remarkable difference from the outer Soviet world was the sound and sight of cheerful smiling people talking freely.

The hors d'oeuvres table sported ice sculptures holding bowls of caviar and reminded me of Sunday night at Hillcrest Country Club, and the members looked the same. After all, many of their parents and grandparents had fled the czar's pogroms. The notable difference was that these middle-aged Russian fat cats wore Levi's instead of custom-tailored attire. The Russians wore their red Levi's tag carefully starched out for plain view and had their Levi's dry-cleaned. High-end dinner followed cocktails and hors d'oeuvres. Between every four or so dinner guests, there were vodka bottles in ice jackets as well as wine bottles. After the vodka of the many-course dinner, our Russian friend said, "Now Amerikanskis, we go drrrrink."

Everyone aboard clapped when the Lufthansa plane took off from Moscow.

# 12

# AMONG THE NATIVES OF NEW YORK

I HAVE MADE a few field trips to New York, some more memorable than others. It seemed like the most foreign place in the world.

Dr. Lou Scarrone had married an heiress and they had a six-story, twenty-five-foot-wide Park Avenue townhouse with frozen dog shit and garbage in the street out front. For that money you could get a whole ranch in Arizona spreading across valleys and mountains. The heiress divorced the doctor, and when he died she got the townhouse back. The rich never let go. But it was Lou's house when I stayed there for a few days.

There must have been windows on the other side of the dark curtains in the drawing room, and there was too much furniture, reminiscent of rented rooms, that hindered a nighttime path to the bathroom. Four floors were just bedroom suites, each sporting a big marble bathroom of a different color. Mine had a bathroom that was all cold black—the shower, bathtub, floor, walls, and toilet. The suite I had was used by the pretender to some small Eastern European throne, but Lou had sent him away to dry out. There were loose stacks of mimeographed pages, that old purple print that soon fades to nothing, extolling his restoration to the throne. The other literature was a bunch of secondhand Mickey Spillane paperbacks.

Dinner guests included sons of famous fathers. One was Adlai Stevenson's. The host and guests ordered out since the hired help didn't do anything and outside was deemed dangerous. Runners brought dinner in duct-taped Styrofoam coolers. Multiwrapped filet mignon, baked potatoes and even butter segregated, and fat-bomb desserts, each condomized in foil.

Sometime later I met the royal pretender and said hello. "You speak to his majesty only when asked." His majesty? That wimp?

I had an appointment at a publisher's and intended to go by taxi but was persuaded to take the subway, much cheaper, my host said. I got lost. Underground you can't tell uptown from downtown, let alone north from south. After a few futile transfers I surfaced somewhere and looked for a taxi. A guy in a large Oldsmobile offered to take me wherever I wanted to go and get whatever I wanted, but I declined. From her bench at the bus stop, a nice old lady said, "White boy, you don't want to be here." Back in the subway I went somewhere else, surfaced, and the taxi to my appointment was expensive.

One really productive field trip, I stayed with friends in a suburb of Aryan lawns and shrubs pruned into bondage. One morning we set out for the warehouse of the Heye Foundation's Museum of the American Indian, where Edward H. Davis's collection of Seri Indian material resided.

My host and guide eventually found the place along the I-95 expressway corridor in the Bronx. Gray, windowless, and multistoried nesting among weedy empty lots, warehouses, and rusted factories. Concertina wire as prevalent as the Eurasian weeds. *Ailanthus* trees were the only thing familiar, those pithy-stemmed slender trees with long lacy leaves and stinky foliage. You see them outside the window along train tracks. Called the Tree of Heaven in its native China and the Garbage Tree here. It's the tree in Betty Smith's *A Tree Grows in Brooklyn*.

I pressed the black bull's-eye button, waited a sensible time, and pressed the bell again with more determination. Eventually the door creaked open a tad, not letting in too much fresh air. A guy in a big black beard who lived inside and looked as if he had never been outdoors greeted us with a grunt. After producing the required papers from the New York office, he let us in, triple-bolting the door behind us and never acknowledging we were expected. As lights flickered on, I was stunned by the greatest assemblage of Native American art and artifacts. More than a million objects acquired by portly banker and early oil heir George Gustav Heye. Whole painted canoes, towering grimacing totem poles, stories painted on tepees and animal skins, weapons, beadwork, basketry, pottery, and more. The endless arts of Native America extracted from vanquished civilizations, cataloged, sorted, and preserved in limbo.

We went up to the floor where the Seri Indian material slumbered in deep wooden shelves and drawers looking like the day it arrived. I was expecting to see the one-man balsa Edward H. Davis bought in 1922 but did not find it. We used Davis's photo of the balsa with Ramón Blanco in our book on Seri Indian ethnobotany. It was one of the last functioning balsas—those sleek reed vessels. The Seris told us that Ramón Blanco was still using his balsa when Davis bought it, and I could go see it "at the university in New York." Davis wrote:

> [I] made a trade with Blanco for his balsa, promising him sufficient lumber, nails and tools, to make a boat large enough to carry five men. . . . [W]hen a Seri man goes in his one-man balsa, he straddles the middle part on his knees resting on seaweed. The craft is light and a good sea craft, but low in the water and the man is wet all the time. He uses a double-bladed paddle, and his harpoon rests against pegs set in the carrizos [reed grass stems that make up the balsa]. A rock which serves to quiet the turtle, two pointed stakes and a heavy fish line complete the equipment. He has to paddle across to the other side of the Infiernillo Channel in that cramped position, say eight or ten miles, very early in the morning, then watch his chance to harpoon. If successful he brings up the turtle, knocks him on the head with the rock, pulls him on the end of the craft by skillful manipulation, cuts slots in the front flippers, puts stakes through and into the carrizo, pinning him firmly in place, then paddles back.

Seri balsas were made from reedgrass (carrizo) stems. Men went to remote waterholes in the interior of Isla Tiburón—the largest island in the Gulf of California—to select and cut the canes in season and carry them to the shore. Each man carried two large bundles suspended from the ends of a mesquite-wood carrying yoke. Leaves were stripped away from the bamboo-like stems, the ends intertwined in a timeless pattern to form long bundles lashed together with mesquite-root rope, and individual bundles were formed into the three large bundles shaping the craft. A large one, ten meters long, would take several days to make and could carry a whole family or six or seven large turtles. In those days large green sea turtles weighed as much as a hundred kilos. The tapered ends of those graceful double-ended crafts rose out of the water when heavily loaded, reducing water drag.

All those cataloged items—some like new, some beat up and well used, perhaps throwaway stuff—frozen from a time when the people lived off the desert and the sea and did not go to a store. Besides, there weren't any stores, and nobody had money. Slit-cane musical instruments: rub them around in your hand on a moonlit night during jackrabbit-mating season and curious jackrabbits hop up close enough to be clubbed. Gaming sticks decorated with Seri blue made from three different plants perhaps related to the species used for the mysterious blue in ancient Mayan murals. A necklace of alternating natural wood and black toasted reed-stem milkweed beads worn by Chico Romero in Davis's photo, like the ones Seri women sell to tourists. All sorts of necklaces of shells and seeds, musical instruments and boxes made from elephant-tree wood decorated with Seri blue, baskets of limberbush strips with red-brown designs from desert ratany root dye, *ikokmolka* fetishes like the one hanging over my computer where I write, and eggshell-ware pottery vessels. Seri women made the largest, thinnest vessels ever, averaging three millimeters thick. Every bit of weight counted when every day men carried those large water-filled jugs suspended in mesquite root nets at each end of a carrying yoke from mountain waterholes to camps at the shore. Examining each item, I wrote furiously, recording what I saw and knew, and what Davis and curators had written on the labels. Again and again objects in my hands related to what the Old People told us and what we observed. I wish there had been time to explore other parts of those vast collections, the magnificence of it all. I felt privileged but also remorse as it was not readily accessible, especially for descendants of the makers. But since then the Heye Foundation went belly up, and it all went to the Smithsonian Museum of the American Indian.

# 13

# ALLEN'S FARM

I WAS ABOUT to head home. New York was way too cold for a desert rat, but Allen Ginsberg invited me to his farm 150 miles north of the city, in Cherry Valley. Said it's warmer upstate. I would have liked to believe him. I had met Allen and his partner Peter Orlovsky in Tucson at Bill and Ann Woodin's house out in the desert east of Tucson. Joseph Wood Krutch and his wife, Marcelle Leguia, were there for dinner, and the Woodins had invited Allen, who was giving a reading at the University of Arizona. Krutch was Ginsberg's literature professor at Columbia. I was planning a field trip to Redfield Canyon the next day, one of the few places in Arizona where the bromeliad *Tillandsia recurvata* occurs—an Ice Age relic from Mexico. Allen was intrigued, so it was arranged for him and Peter to come along. Allen reluctantly agreed to 7:30 a.m. after I tried for 6:00.

Leaving the graded dirt road at Redington, I headed my old white Scout in four-wheel low-range up the rocky road into the west face of the Galiuro Mountains. Peter suddenly demanded a stop to "give the bunny rabbit an apple." Allen said, "But Peter, we don't have any apples." About halfway I stopped to botanize in a rocky arroyo, to look for some rare spring plants in flower. Peter pointed his finger at the arroyo and started yelling that there was a dead man. Allen told him it was just a dead cactus.

Two hours later at the end of the road, at the canyon edge, an eroded rock trail leads a few hundred feet steep down to the riparian canyon. The terrain was rougher than Allen had imagined. Said he had hip trouble and stayed by the car while Peter and I clambered down the cliffside. Peter saw only one more dead man and I told him it was just a dead log. Later in the day I showed Allen how the stamens of a cholla flower close up when touched, like when a cactus bee stimulates the stamens after diving into the center of the flower to get its nectar reward. Enclosed by stamens, the bee has to push its way out through pollen-laden anthers. In Phoenix a few days later, a reporter was goading Allen about his lifestyle, and Allen asked if that was any more perverse than a bee sucking off a cactus flower and getting embraced by flowering sex parts to ensure fertilization. The reporter thought Allen was being a smart-ass and punched him in the nose.

So instead of going home, I met Allen on a cold morning at the New York Port Authority bus terminal. Briefcased commuters scurried past the homeless slumped at the walls. The cops didn't hassle them as long they stayed along the walls of that great cathedraloid station. I thought about those sci-fi books describing the grim future of Gotham City, as I had ten days earlier in Washington seeing homeless women sleeping on warm air grates along the side of the United States Supreme Court.

The sky was low and gray and the road to the farm snowed in, so we had a long sloshing wet walk through knee-deep snow. After changing to something dry and drinking some warm concoction, I finally stopped shivering, but that night I started coughing. The cozy old farmhouse was warm but too small for so many people leeching off mother hen Allen. Furnishings were completed with oil-clothed kitchen table, blotchy linoleum, and sagging couches. Halfway through dinner, Lafcadio, the only one of Peter's three brothers not yet institutionalized, launched into a dissertation on salt mining when asked to pass the salt. Someone said it was time to put him back on his meds. The dissertation was ended by asking him to take out the garbage. Later someone realized they had asked him to take out the garbage but forgotten to tell him to come back in. Poor Lafcadio standing out in the cold until someone told him to come in.

The coughing ramped up without relief, and I was apprehensive because of a collapsed lung last summer diving off the desert coast at Toliara in Madagascar searching for sea grasses. A repeat would not be good. I asked Allen if he could get some cough medicine, but that was not an option in Cherry Valley because Peter had threatened to beat up the only doctor when he wouldn't write a scrip for speed. Peter at six-two was built like an ox. I called Dr. Lou Scarrone, whom I had stayed with in New York, and he phoned the local pharmacy, vowing I was his patient and it was vitally important. More snow fell and a kind soul snowshoed to town to get my prescription. I had sheepish thoughts of Iditarod heroics. Allen casually said to hide the cough syrup. One good slug worked magic. When the effects wore off it was time for another dose, but the bottle was empty. Junkie poet Ray Bremser was sprawled out in the next room singing incoherently. From the other upstairs room, Peter and his girlfriend were publicly busy with love.

Allen's phone rang mornings from London and nights from Tokyo. It was a reporter from London, and Allen wondered what to tell him that would be something unusual. I said tell him about the plight of the whales, the Japanese won't stop killing them, Iceland and Norway too, and they weren't the only ones. I provided an earful about the shame of whaling, and Allen added poignant words as only a great poet could. I thought how powerful it would be if he wrote poetry about the planet. The phone call was a success. Whales broke into the international press big time.

# 14

# WHALE TALES

I THOUGHT ABOUT the whale issues at the first United Nations Conference on the Human Environment (1972). My official identification entry badge said "Friends of the Sea Otter" and "Defenders of Wildlife" (then just a tiny organization run by a steel-nerved old lady who founded the nonprofit using her basement mimeograph, before George Whittell Jr. bequeathed millions). Other participants included heads of state, cabinet-level ministers, famous scientists, and NGO bigwigs. At the red carpet Stockholm Royal Palace reception, I gave a Japanese gentleman a talking to about whales.

"But Dr. Felger, you don't understand."

"What is it I don't understand? If you kill the last whale there won't be any for your children and grandchildren's grandchildren."

"But Dr. Felger, you don't understand."

"A great nation like Japan should not bear the shame of wiping out the greatest animals ever on earth."

"But Dr. Felger, you don't understand."

We tangled round and round, over and over. Later I was severely admonished for talking rough to the Japanese minister of finance, the most powerful man in Japan. Margaret Mead complimented me. Like an elfin Brunhilde, she stomped around stamping her big crooked wooden staff, coercing recalcitrant bureaucrats to do the right thing. I would have liked to have given the minister Michael McClure's "For the Death of 100 Whales."

Stockholm authorities had set up facilities for the vociferous alternative conference inconveniently out of town at Skarpnäck Field (some called it ScrapNeck), but the protesters filtered downtown anyway, even to the royal palace. Joan McIntyre and the Hog Farm Commune's rousing whale march got a lot of press. Michael McClure said Joan would have liked killer whales to be vegetarians, certainly not munching sea otters. (Jim Malusa told me he had a similar conversation with a vegetarian about the Inuit: How come they weren't vegetarians?)

Gary Snyder read Mother Earth ecopoetry, and official U.S. delegates objected to everything and tried to stifle Vietnam horror and protest. I got nowhere on sea turtle conservation, let alone endangered plants. I was a few years too early. Stewart Brand, Sterling Bunnell, Joan McIntyre, Michael McClure, Joanna McClure, David Padwa, Melissa Savage, Gary, and some Hopis: We all stayed at a drab little hotel that had a pay phone in the miniature lobby. The Hopis put tiny Swedish dimes in the phone and talked Hopi to Arizona for hours on end. Someone said Stewart was paying for the calls.

One evening I followed Margaret Mead into a Gamla Stan (Old Town) restaurant, ducking through the low doorway below street level once aboveground. I wore my Friends of the Sea Otter and Defenders of Wildlife badges and sat across from Nixon's secretary of the interior Wally Hickel. Conversation was more than lively since Friends and Defenders were suing Hickel. Although I didn't like a lot of his policies, I came to admire the man for his anti–Vietnam War pronouncements, galvanized by the Kent State shootings. Nixon promptly fired him.

Sterling discovered a Mexican restaurant, and taco intrigue in Stockholm was not ignored. Red carpets, maître d', tinkling crystal glasses, and blue-eyed Swedes trying to sing corridos. The tacos were crepes filled with unfortunate creamy concoctions and they weren't cheap.

After the big conference, we took a train ride to Uppsala to pay homage to Linnaeus's garden and home, passing Scandinavian forests where the trees line up in rows. The great man's desk was surprisingly simple for one who led his troops on field trips announced and ended by forty French horns and kettledrums. The popular excursions started from the Royal Botanic Garden. Linnaeus spoke in Latin, recorded by a scribe who gave the pronouncements to an assistant, who told the troops in French what the great man said. His very popular sexual classification of plants brought the ire of the church.

On the way home Michael and Joanne and Sterling and I stopped over in Iceland. The oldest democracy and freest press in the world, the oldest vegetarian restaurant, midsummer midnight sun, the first country to achieve 100 percent literacy, where farmers write poetry, and still they kill whales. We had to witness the geothermal Hvalur whale-processing factory, deep in a green-walled fjord. Summer workers in rubber boots sloshed across concrete floors in blood and gore, flensing fluted limp white whale bellies. Iceland hunts endangered whales, selling the meat to Japan and to feed mink fur farms. New Yorkers need mink coats.

I stayed on in Reykjavik, eager to see more of such an exotic ecosystem. My hosts wondered why American houses were so dirty; they cleaned the bathtub with a toothbrush. Near midnight I helped a biology teacher haul his small boat down to the Reykjavik dock. Passing puffins, skuas, and kittiwakes on sea cliffs, we headed to where land disappeared as the sun dipped below the horizon, but it never got dark. We fished haddock and hauled the silvery catch to his house, swinging open heavy wooden doors to the underground icehouse. Like so many Icelandic families, they had enough to last through winter. He didn't answer when I asked why they still kill whales.

Whale flensing at the Hvalur whale-processing factory, Iceland, June 1972. Richard Felger archive.

# 15

# NOTES FOR NEXT TIME

EVER SINCE CHILDHOOD, it was obvious that the world is one superorganism. When I read James Lovelock and Lynn Margulis's Gaia theory, it hardly seemed like a theory but a description of the obvious. My mind skipped to the interrelated trends of warfare and escalating population, later galvanized by 9/11.

If we are so smart, why do we still have war? Why are millions of people suffering unspeakable abuses? Why are animals and plants disappearing? Why is your government stockpiling nukes to wipe us off the planet?

If there is adaptive value to consciousness, it is to do enough before it is too late. You might say the problem is too big. Looking back, seems like the debacles in Iraq set the stage for WWIII. Trying to predict the future? Increasing population will lead to bigger disasters, ultimate global warming, pandemics, bigger weapons, and eventually someone will use them. Bang bang you're dead; the turtles and the whales and all the fish are dead. Everything is dead. Nuclear winter, nuclear summer, nuclear death. It's not science fiction; it's just out of sight beyond our horizon. But this scene is based on present conditions continuing, and we don't know the future. Perhaps an addictive male birth-control pill will be in the hands of women.

Crank fast back past the Industrial Revolution, back past the Renaissance, Rome, and King Tut, to Ice Age mastodons, cave bears, and lions in the place we now call England, and keep cranking to prehumans running around the Serengeti Plains making hand axes—handles not yet invented, though fire has been tamed.

They used the energy of the day, plants, animals, and then fire to release tree energy. Scroll up through the millennia to the written word and harnessing the wind with sails to conquer far from home. Whale oil and oil of geologic time; entropy to balance human negative entropy. No single animal species ever so weighty as humans. Population sneaking up until two billion requires balance of cosmic energy release: Trinity, Hiroshima, Eniwetok, Chernobyl, and the list goes on. Faster and faster doubling and redoubling artificial intelligence and communication and consumption into the balance.

You told me eight, ten billion would be impossible. Is there an adaptive value for consciousness? What can we really do? The only thing I know is to keep on botanizing and writing.

PART II

# FROM FRIENDS AND COLLEAGUES

16

# A PRETTY INTENSE FELLOW

Thomas Bowen

IT WAS, OF COURSE, Becky Moser who introduced us. It was 1967 in the Seri (Comcaac) village of Desemboque. I was doing an archaeological survey of the Seri coast and Becky and Ed were helping me with Seri ethnographic data. "Have you met Dick Felger yet?" Becky asked. (Richard was Dick back then.) "He's studying the coastal plants and he's in town right now." She lowered her voice and added, "He's a pretty intense fellow."

Richard was just leaving, so we didn't say much, but I knew that anyone working in the Seri area back then was someone worth knowing.

Late one afternoon two years later, my friend Bennie and I were walking across the University of Colorado campus, where I had just finished my graduate work, when a voice from behind said, "Hello, Tom Bowen." I turned, and there was Dick Felger.

"What are you doing here?" I blurted out.

"I used to teach here."

He said he had just walked out of the World Affairs Conference (a major annual event at the university), where he was an invited speaker. "I tried to tell those arrogant dignitaries that they're wasting their time debating monetary policy and technical assistance for underdeveloped countries when the real problem is uncontrolled population growth. They didn't want to hear the truth." It was the first I knew of Richard's lifelong involvement with this critical issue.

I introduced Bennie and said we were going to pick up a pizza and head over to my place and would he join us? "Do you have a friend you can bring?" I asked.

He didn't reply but took a quick look around. Just then a pleasant-looking coed happened to be walking by, and Richard moved quickly to intercept her.

"Excuse me, Miss, my name is Dick Felger—what's your name?"

Caught completely off guard, she replied, "Uh, Cynthia."

"Cynthia," he repeated, "What a beautiful name. I hope you don't let people call you Cindy."

Wow, this guy is good, I thought, as Richard poured on the charm. A short while later the four of us met up, pizzas in hand, at the sagging one-room converted garage I was living in. I don't remember much of the evening except that Richard was delightful company; we probably talked way too much about the Seris, and Cynthia was bewildered by the whole scene.

After that evening, Richard returned to his job as senior curator at the Los Angeles County Museum, and a few months after that, I began teaching at California State University, Fresno. I saw him a few years later at a conference in Hermosillo, and Richard came through Fresno once and stayed overnight at the aging little tract house I was renting. At the front door he pointed to the tall clump of decorative reedgrass—known locally as "bamboo"—and asked, "Did you know you could make a Seri balsa out of those? They're closely related to the *Phragmites* reedgrass of the Seri coast." I had no idea.

Over the next several years, we corresponded occasionally when one of us needed information. One letter, though, about 1975 was very strange. Richard and two colleagues had learned that sea turtles hibernate in the shallow channel between the Seri coast and Isla Tiburón, and that the Seris had known all about it long before scientists had. He wanted my advice on where they should publish their paper about this discovery.

Why ask me? I wondered—I'm not a biologist. I knew nothing about sea turtles and next to nothing about journals outside anthropology. I reluctantly suggested *Kiva*, a regional journal of anthropology and history that had recently published several Seri articles, including three of Richard and Becky's early ethnobotany papers. Needless to say, I was thrilled for Richard when the sea turtle paper appeared a year later in *Science*, the most prestigious American scientific journal. Congratulations, Richard!

We met up again in 1978 at the Tucson meeting of the Society for American Archaeology. By then I was no longer surprised to see him turn up just about anywhere, as I knew that his intellectual world extended far beyond botany. I was hanging out at the conference with two friends from my department at Fresno, and Richard invited the three of us up to his little bungalow in the foothills above Tucson for tequila and tonics. It was a bit tricky getting into the house because there was an ungainly hole and a mess of big rocks and gravel in the only flat spot right next to the entry where he had a swimming pool under construction. Once inside, Richard was an ever-so-gracious host and thoroughly enjoyed himself as he served us our drinks in chemistry beakers! I thought that was so cool that when I got home, I ordered a set of beakers and for years used them as beer glasses. Today it's an occasional beer from a bottle, but I still drink my morning orange juice from a dedicated beaker.

Richard and I genuinely became friends when I spent the 1986–87 year on sabbatical in Tucson at the Arizona State Museum. He and Becky had just published their monumental Seri ethnobotany book *People of the Desert and Sea*, and I was deeply immersed in research on the extinct Isla San Esteban band of Seris, so we had a lot to talk about. Moreover, we were both single, and neither of us was bound by a nine-to-five job, so it was probably inevitable that we began to pal around together. We often went out to lunch—the Blue Willow on Campbell Avenue was a favorite—and occasionally dinner at one of Tucson's many superb Mexican restaurants.

We both liked to hike, and sometimes we drove up to the trailheads high in mountains above Tucson to escape the heat. Richard usually volunteered to drive—he had a light blue Mercedes-Benz sports car convertible that he was clearly fond of. And it was not his first Mercedes, he told me. He had once owned a Mercedes-Benz 300SL Gullwing—a super-fast high-performance sports car with doors that opened upward—like a bird's wings flapping—instead of out. It was also rare. One day he launched into a tale involving the only other Gullwing in Arizona, which was owned by a rich British countess who had come to the United States and lived in Tucson. Apparently they had an uneasy competitive relationship based on whose car was the fastest. It finally ended in a showdown at dawn on the Avra Valley highway, west of Tucson.

I wish I had paid more attention because I don't remember the outcome. At the time, the story seemed utterly unbelievable—like an updated Old West gunfight—and I thought he must be putting me on. Or was he? As I got to know Richard, I realized there were aspects to his character that I never would have imagined. Maybe it really was true.

Richard, barefoot with pants rolled up to his knees, duplicates Carl Lumholtz's 1910 photo of Salina Grande, December 1967. Photo by Thomas Bowen.

Richard also had a four-wheel-drive Chevrolet pickup truck with a camper shell that he used for fieldwork. It was the vehicle we used that December for a five-day trip around Bahía de Adair at the head of the Gulf of California. Richard and three colleagues were making a comprehensive study of the region's *pozos* (freshwater springs), which were first documented in 1910 by Carl Lumholtz. Richard's role was to inventory the flora and assess the changes since then. He invited me along to document the archaeology and, I suspect, for company.

Unlike some of Richard's legendary field trips, with their frenetic pace and twenty-hour days, this one was relaxed and thoroughly enjoyable. At each *pozo*, he did his work and I did mine, with plenty of time for a nice camp dinner and a good night's sleep. The exciting part came at the end, when we drove from Pozo la Salina (Salina Grande) down to the shore, deflated the tires, and drove the fourteen miles along the sandy beach to the road head to El Golfo de Santa Clara. You have to do this at low tide because at high tide the beach is completely under water, and the waves lap up against the bluffs directly behind it. If you mess up on the timing, you're in deep, uh, water.

Richard and I both grew up with classical music, so we sometimes took in concerts when touring musicians came through Tucson. These were world-class performers, and one was the great violin virtuoso Isaac Stern. Despite his fame, I confess I've never much liked solo violin, and so while Richard sat there in rapt attention, I fidgeted through the entire program wishing I were somewhere else.

One day Richard told me he had to have a hernia repaired, and he asked me to come by while he was convalescing. I dropped by at his house the day after the surgery and quickly concluded that he didn't really need me. He was stretched out on the sofa under a blanket, a table beside him with food, water, and pain pills, and two doting women friends hovering over him and responding to his every groan. After I expressed my sympathy and asked the obligatory, "Is there anything I can get for you?" I quietly made my exit, thinking he sure did know how to be an invalid.

My office at the museum was in a bleak, cavernous, windowless room, and I appreciated anybody who came by for a visit. One Saturday afternoon I was sitting at my desk, bored, when Richard walked in with a heavy pack on his back.

"You've got to see this," he said. "This is incredibly neat." He pulled off his pack and extracted a large ungainly fourteen-pound plastic box. "It's a laptop computer! Everything you need is in this case. You can plug it in, but it can run for an hour on batteries, so you can take it anywhere!"

He was a kid with a new toy as he opened the case and showed me how it worked. I'm not sure, but it may even have had a hard drive—super high tech for 1987. "You've got to get one," he said. "Forget about those mainframes and clunky desktops." And I did. A year later my first computer was a laptop, and I've used nothing but laptops ever since.

After that year in Tucson, Richard and I kept in frequent contact. He was a great communicator, and I could expect a phone call from him about every two or three weeks. Sometimes he called for specific information for something he was working on or to discuss something he was thinking about. Other times it was just to chat. Either way, these were lively conversations spiked by Richard's dry wit and irreverent view of life. And sometimes I couldn't figure out whether he was serious or just leading me on. Although our telephone relationship sometimes seemed lopsided—with him calling me more than I called him—it really wasn't, because when he called I often had as many questions for him as he had for me.

We hardly ever corresponded by letter, but we did use the postal service to exchange manuscripts that needed a critical eye. This *was* a lopsided relationship because the ones I sent him usually needed a lot more attention than those he sent me, and he gave it generously and diplomatically. Richard also occasionally sent me published papers that he considered of special importance. One of them proposed *Distichlis palmeri* (nipa, or Palmer grass), a salt-water grain-bearing plant, as a potential crop species that could be cultivated to feed an ever-increasing human population. And he sent a few manuscripts of a more personal nature, including accounts of some of his wacky field adventures that he was gradually assembling into a collection he called *10,000 Years of Field Notes*. He knew I would appreciate these stories because we had often commiserated about how few scientists ever reveal what fieldwork in remote places is *really* like.

Whenever I came to Tucson, we would get together, often for lunch at his house. On one of those occasions, after finishing our sandwiches of avocados and vegetables from his garden, he suddenly and rather emotionally brought up the sea turtle paper he had published in *Science*, confronting me with my failure to recognize the paper's significance and my naïve suggestion that he publish it in *Kiva*. I had long since forgotten the incident and had no idea that he had been deeply offended by it and still harbored resentment some twenty-five years later. Apparently, raising the issue that day brought closure, and he never said anything more about it.

On another occasion he showed me a framed photo of his mother and her second husband, dressed in formal wear, at a society event. It was then that I realized that I had never heard him say anything about his family or his childhood. Maybe I shouldn't have been surprised—I don't think I had ever mentioned mine. Sometime later I learned that he had a brother when he told me he had to go to Los Angeles for the funeral. Much later, at a restaurant, he spontaneously related a fascinating story about his grandfather, who had established a general store right on the Mexican border in Naco, Arizona, and the chaos he had experienced during the Mexican Revolution.

By the late 1990s Richard and Silke were together, and my wife, Marty Brace, and I were retired and living in Lander, Wyoming. She and I often traveled to Tucson and made a point of getting together with Richard and Silke. On one of those occasions, Marty and I treated ourselves to a week at the renowned Arizona Inn and invited Richard and Silke to join us there for dinner. Richard and Silke were late, but no matter—we were enjoying the ambiance of the restaurant, which was a hangout for movie and TV people who often stayed at the Inn. In fact, at the table next to us were the actors Ron Glass and James Brolin, who flashed Marty a nice smile.

Half an hour had passed when a visibly upset desk manager found us and said that she had had a "very difficult" phone conversation with a Mr. Felger, who wanted her to tell their guests named Bowen that they would be late. Uh-oh, I thought, knowing how difficult Richard could be at times. "What happened?" I asked. She said she told Mr. Felger that they had no one named Bowen registered, but he insisted repeatedly that they did. After several heated rounds of "Yes you do" and "No we don't," Mr. Felger had asked, "Well, do you have anyone registered from Wyoming?" Yes they did, but it was not a Mr. Bowen—it was a "Mr." Marty Brace!

Marty, of course, was Ms. not Mr., and Richard and the desk manager finally realized that it was she who had registered us. So the mystery was solved, Marty and I apologized profusely to the unhappy desk manager, Richard and Silke duly arrived, and we had a sumptuous dinner with the stars.

We also had some very special dinners at Richard and Silke's house. One in particular, sometime around 2000, stands out for the way the conversation evolved and what it led to. Somehow the subject of Ike Russell, the legendary Tucson bush pilot, came up, and Richard—who was a splendid raconteur—began regaling us with stories of his madcap adventures flying with Ike. I had had my own adventures in Ike's airplane, and I had heard others tell of theirs. But what struck me, listening to Richard's tales of wild times and near disaster, was that these "Ike Russell Stories," as they were known, made up an important body of regional folk tales that needed to be recorded before they were lost forever. Thanks to that dinner, a large collection of these stories appeared two years later as a book under the title *Backcountry Pilot*.

About that time, the internet was making its appearance, and the miracle of email was looming on the horizon. I was just getting the hang of dial-up connections when I got a call from Richard, who said, "This is the last phone call you're ever going to get from me. From now on it's email or nothing!"

And he wasn't kidding. At first I did call him a few times, but he was so ornery about it that I soon gave up. Email it was.

Richard took to email like a duck to water. He gleefully adopted some of the techno-nerd conventions, such as using the contractions "yr" for "your" and "shd" for "should," and dispensing with capitalization and sometimes even punctuation. Like so many users, email freed Richard from the minimal mental planning and organization that letters and even phone calls required. His emails were spontaneous outbursts and often came in rapid-fire salvos of four or five (or ten!) as he kept thinking of things he wanted to add to the emails he had just sent. The instant communication that email provided also seemed to have accelerated his requests for action—"plse plse do it right now—i hve to hve it by monday." But trading emails with Richard was also great fun. He loved the word play it enabled, and the immediacy of response made it almost like talking with him.

Richard was also quick to adopt serious internet tools he found useful. One day we agreed to meet at the University of Arizona Herbarium to go to lunch, but when I arrived he excitedly exclaimed, "Tom, you've got to see this—this is an amazing development."

"This" was Google Earth, and for the next half hour he demonstrated some of the things you could do with it. He showed me exactly where on Isla Tiburón a Seri woman had taken him and Cathy Moser to see a rare stand of wild tepary beans, how the program precisely pinpointed the location with GPS coordinates, and how their route to the stand could be plotted and the distance measured in any units. That convinced me that Google Earth was an essential tool for anyone who did fieldwork. I hate to think how long it would have taken me to discover it had it not been for Richard.

After Richard and Silke moved to Silver City in 2007, Marty and I saw Richard only once, in 2011, when Silke was out of town. He joined us for dinner at the pleasant country inn he had recommended, and he was his usual gracious self. One morning he and I went for a long walk alongside a slowly flowing creek and talked at length about things both personal and global. It was a subdued conversation, a departure from times past, conditioned perhaps by our ever-advancing years and humanity's failed opportunities.

After that visit, our emails tapered off some, though about 2018 he sent me the manuscript of his monumental Yaqui (Yoeme) ethnobiology book and asked for comments ("i must must hve them by friday"). In October 2020, not having heard from him in a while, I emailed him with a couple of inconsequential questions and a little light chatter. He replied, "cancer has come back—does not look good."

I was stunned—I had no idea he was sick. I wrote him a short message but got no reply. A short time later Ben Wilder emailed me to say that Richard had died.

People like Richard don't come along very often. He was a consummate scholar, a cherished colleague, an uncompromising conservationist and humanitarian, a gifted storyteller, and a person who never lost his quirky sense of humor. But for me he was, most of all, a wonderful friend.

17

# THE PLANT GUY WHO DROPPED IN

Cathy Moser Marlett

WHEN I WAS a child growing up in the Seri village of Desemboque, Sonora, my playmates were Seri kids, and our playground was the seashore and the seemingly endless desert behind town. We scoured the shore during extreme low tides, collecting shells and turning over rocks in search of octopuses, while our desert forays included interacting with the array of desert plants we found. We sampled many of them, enjoying the fleeting sweet taste we sucked from the bright red-orange flowers of the ocotillo (*xomjeeziz*, as my playmates knew it), or eating the mouth-puckering fruit of the jumping cholla (*tootjöc*), or sampling the bulbs of the blue sand lily (*caal oohit*) that we scrounged from the dunes. Until Richard Felger appeared in my life, I had no idea that every plant in my world also had a Latin name, and that these were considered by botanists to be their *real* names. This new dimension of oddly spelled names soon became part of the personality of the plants I knew.

I was about eleven when Dick (as we knew him then) stepped through the back door of our house, large plastic collecting bags in hand, and without a howdy and nice to meet you abruptly asked where he could press his plants. My nonplussed mother obligingly showed him our kitchen table instead of the door—and there began many years of a very special collaboration, culminating in the book *People of the Desert and Sea: Ethnobotany of the Seri Indians*. For this comprehensive study, Richard supplied the scientific botanical expertise, and my mother—Becky, as she was known—provided the Seri knowledge of plants learned through conversations and interviews with the Seri people in their own language. As the project developed, Richard, in Tucson, would send down pages labeled "homework for Becky," with lists of questions: Where was a plant found, how was it collected, was it medicinal, was it eaten (if so, how, and when), and any other information that could be added. A single question often opened fascinating paths of Seri knowledge, and the whole thing got more and more complicated and ended up being about much more than just plants. Richard later recalled with much humor (and his characteristic snicker) that on that first collecting trip, when he showed up at our house and saw the Seris, he was somewhat annoyed and wondered who these people were and what they were doing in his plant-collecting area! I think they grew to enjoy his visits and interactions with him, and especially liked to see how their botanical expertise was becoming appreciated in the outside world. As frequent visitors usually got dubbed with a Seri name reflecting some notable characteristic, Richard quickly became known as Hehe Iyat Ctamtim, literally, "plant tip picker," or, as it was given colloquially, "The Guy Who Cuts the Tops Off Plants."

It wasn't unusual for field researchers to drop in for a meal or an overnight stay at our house, which had the only hot shower in town. The conversations were always interesting, at times providing me with different perspectives while exploring my desert and ocean playground. There would often be something unusual happening at our long kitchen table, like Richard or others preparing lizards for preservation in formalin (and me being given glass vials for my own collections), all the while being observed by curious Seris. Thanks to these visits, I became adept at collecting insects, learning how to mount and label them, and creatively preserving other creatures I found, like the freshly dead shore bird that I cleaned and stuffed and hung from the ceiling in my bedroom, where it stayed for years.

Richard's visits were some of my favorites because they provided wonderful opportunities to escape from my mornings of home schooling. There was the usual plant pressing at the kitchen table, but there were also trips into the desert to locate a plant he had heard about, and sometimes I got to go along on these bumpy rides, squashed into the back of his vehicle. I was rather awed by his intense and singular passion for plants, and I always had mighty hopes of impressing him by showing him a plant unknown to science (it never happened). Instead, on one occasion, I rather did the opposite. We had located an

The Mosers' famous kitchen table in Desemboque, March 1978. At left, Ike Russell converses with his daughter-in-law Susan Randolph, while Doris Potwin (*front left*) and Jean Russell (*front right*) listen in. At the end of the table, Richard and Becky work on their ethnobotany notes. Image courtesy of Cathy Moser Marlett from *Backcountry Pilot*, edited by Thomas Bowen © 2002 The Arizona Board of Regents. Reprinted with permission of the University of Arizona Press.

unusual slender-stemmed cactus growing tangled in a large bush, and he was quite interested and recorded it. As he walked away, for some strange reason I grabbed it and broke its stem. I was completely mortified, certain that he would turn around right then or revisit it some day and find a dead rare plant. I didn't confess, and I'm sure he never went back. On another excursion to a long spit bordering an estero, we passed by a small cactus I didn't recognize, and I drew Richard's attention to it. After a quick glance, he muttered, "I didn't see that!" and walked on. I still don't know what he meant by that curious remark, although I can imagine he was avoiding tortuous last-minute edits to the ethnobotany book.

Richard taught me to be aware of my impact on other creatures. Once when I was, as usual, turning over shore rocks looking for octopuses, he somewhat pointedly told me to return the rocks to their original position because of the ecosystems beneath them—that was a new word for me, and something that had never mattered to us kids. He didn't relate easily to children and was at times blunt and somewhat stern, but I think I accepted that because I respected him and wanted to learn.

The best adventures were being invited along on overnight trips to the nearby islands with Richard and another passenger in Ike Russell's four-seater plane. These were memorable times, of swimming near the sea lions of Isla Patos; of hiking the interior of Isla Tiburón with Rosa Flores, a Seri woman who was a major contributor to the ethnobotany book, to locate a small patch of wild tepary bean (haap, *Phaseolus* sp.); and of visiting Xapij (Sauzal), a permanent water source lined with reedgrass on the south end of the island, and then flying home over isolated beaches with large numbers of sharks swimming far below. There was so much to see!

In later years I was away at high school and college in the United States, coming home only during Christmas break and in the early autumn, after we spent most of the very hot summers elsewhere. I missed the many research activities around Desemboque during those years, but my parents described them in frequent and detailed letters, which I now treasure. Some of them recount helping Richard as he investigated the Seri use of eelgrass and the then generally unknown phenomenon of sea turtle winter dormancy, which came to light because of how the turtles were hunted by the Seris. I recently found a letter from my father that told how happy he was to obtain a seaweed sample from the carapace of one of these turtles for Richard, no doubt part of the often-requested "homework." Because of his work with Seri consultants, my father played a key role in making the winter dormancy known to science. However, before Richard and two co-authors submitted an article about this to the journal *Science*, my father pulled his name as a co-author and withdrew his name from all other mention. Seri hunting of dormant sea turtles was a sensitive issue, and my father was concerned that association of his name with the article would jeopardize his relationship with the Seris. Although Richard supported his decision, I regret that my father's contribution never received credit.

By the time I was studying biology in college, my relationship with Richard had changed. During the spring quarter of my junior year, I conducted an off-campus independent study project at Estero Sargento, a large estuary an hour south of Desemboque, and Richard served as an adviser. As a child, I knew the estero as a favorite place to go to get large blue swimming crabs for a dinner feast at home. But now it was a place to investigate, learning from Richard such things as making transects, and to collect and record what I observed about the estuary plants and faunal life. One of the most memorable events was a nighttime foray with Richard in a Seri boat into the deepest channels during the incoming tide, with a gas lantern hung at the bow, illuminating the estuary floor where fish swam and scores of Cortez garden eels swayed in the current, partially emerged from their burrows. That study was a time of interweaving my early experiences with science, giving me the excitement of learning on my own. I loved it. It was also where my deeper interest in mollusks began, and with Richard's encouragement, I began to investigate what the Seri people knew about marine mollusks. (Richard later told me that if he hadn't chosen botany, he would have developed his other interest, malacology.) Of course, I was greatly inspired by his and my mother's finished ethnobotany book. When Richard saw my ever growing list of mollusks with Seri names, he strongly suggested that I publish it right then. But I wanted it to be more than just a record of names, so I continued research and writing during the next few years of visiting the Seris. Richard was always an enthusiastic mentor, but he no doubt recognized that research that was so much fun for me might not see a conclusion, and he eventually urged me to call it quits and publish. He sent an email to the University of Arizona Press, copying me, and saying, "You have to publish this book." And they did—*Shells on a Desert Shore*. I owe this and much more to Richard.

I remember on a short car ride with him in Tucson in what was maybe our last in-person conversation, reminiscing about our years of connection with the Seris. He told me that he thought of my parents as "real" Christians, and that he had loved them. It was a relationship that had grown over a long span of time, and it was not always without difficulties, as Richard was rather passionate about his work and sometimes oblivious to others. A nonreligious scientist and protestant Bible translator-linguists were a seemingly odd match, but it was a lasting friendship valued by each. I treasure having had Richard as a part of my life as well, and certainly, the world of Sonoran Desert plants has lost an incredibly productive, intense, and unique man.

18

# THE BEST FIELD TEACHER

Gary Paul Nabhan

RICHARD STEPHEN FELGER was never known to be a great classroom professor, though he did try it once or twice.[1] And yet, dozens of my desert friends and most esteemed colleagues would agree that he was among the best and most beloved field teachers of natural history, desert botany, and ethnobotany that they ever had. He would literally pull you into being part of a "plant family" by offering the most interesting, whimsical, and wonderful takes on each particular plant we encountered. Sometimes he would let out a nearly orgasmic moan, yelp, or holler when he encountered a new species or a disjunct one hundreds of miles out of its range.

The desert legion of botanists, zoologists, ecologists, anthropologists, ethnobotanists, and geographers that he mentored reads like a who's who of Borderlands environmental sciences over a half century's time.

And yet all of us had to learn to accept not only that Richard's botanical brilliance was far beyond ours, but that his eccentricities were legion and legendary as well. At the risk of being irreverent, I will try to capture *the whole enchilada* of Richard Felger, because he was hilariously irreverent. He once told two somewhat stuffy Harvard friends of ours that he was sorry that they had to

1. Adapted from *Water in the Desert: A Pilgrimage* (Milkweed Editions, 2026) by Gary Paul Nabhan. Used by permission of author and publisher.

live back in that temperate forest, where there was only a monotonous "insipid green" covering the ground, rather than being immersed in the rainbow pallet of the Sonoran Desert. In fact, the only two times I heard him speak with reverence about someone was when the Beat poet Michael McClure showed up at his door, and when he spoke of hosting the great ethnobotanist Richard Evan Schultes out on Isla Tiburón. The rest of his time, the serendipity of whimsy guided his soul into both ecstasies and disasters. I cannot honor this great man unless I speak to both those moments of joy and the moments of chaos. I "survived" forty years of co-authoring a dozen scientific or popular articles and book chapters with Richard only by coming to accept the whole enchilada, not just the glistening cheese and cilantro on top. He was the greatest botanist the Sonoran Desert has ever known, but he was also as thorny as a teddy-bear cholla cactus population on a volcanic slope in the Sierra Pinacate.

I happened to be lucky enough to be reading Richard's ethnobotany articles by the time I turned twenty-one, and to have savored his desert poems by the age of twenty-five. When ecologist Ron Pulliam spotted me reading Richard's articles at the ranch house dorm we shared at the Research Ranch in Elgin, Arizona, he encouraged me to go talk to him.

"I think Richard is looking for a field research assistant for an NSF grant he just received to finish up his book with Becky Moser on Seri ethnobotany. You should go to see him. I can give you his phone number."

"Where does he live?"

"Right next to where I live in Esperero Canyon. That's up by Sabino Canyon in the Santa Catalina. My wife, Janice, and I have been living in Ed Abbey's house, and we've finally purchased it. Richard lives with a wonderful musician, Mahina Drees, about sixty yards from our place. I'll draw you a map in case you call him, and he has time to see you."

I called the number, and Mahina answered. Wonderful cheery voice.

"Let me look for Richard. He may be out front in the desert watering some cacti he transplanted."

Five minutes later, Richard came to the phone. Yes, I could come up. Saturday morning might be best.

"We'll talk when you get up here, but yes, I do have a job opening. I need to get back to the plants."

When I got back to the dinner table where Ron Pulliam was sitting with a University of Colorado zoologist named Rob, I told them that I was thrilled and hopeful that I had a shot at working with Richard.

That stopped Rob in his tracks. He put his fork down on his plate and said, "Did you say *Richard Felger*? I haven't heard his name in years."

Ron and I immediately blurted out, "How do you know Richard?"

> He taught briefly at the University of Colorado. *Very briefly*. I took his botany class, I think even helping like a teaching assistant for lab setups. . . . But he was so homesick for the desert that he'd leave Boulder as soon as his Thursday class was out, then fly down to Tucson where a guy named Ike Russell would personally fly him down to the coast of Sonora for his fieldwork with Indians there. Then he'd get back just before class on Tuesday. We hardly ever saw him. But one time when he heard I was going up to the Rocky Mountain Lab for two weeks to live trap alpine mammals, he asked if he could use my log cabin in the mountains above Boulder for a party for some of his friends coming in. I consented, but boy, was that a mistake. When I returned to Boulder, my cabin was in ashes. Ashes. It had burnt down during the party. When I called Richard to ask what happened, he simply said he was sorry, but he had left the all-night party early because he was tired and had to get ready for a flight down to the desert. He only heard that it burnt down after he returned and wasn't sure who was responsible! He said to let him know if I needed any information to help him get insurance for it!

Fortunately for many of us, Richard soon returned to the desert to live full time, rather than teaching in a faraway university. Thus began one of the most bizarre relationships I've ever had with another human being, if Richard ever was *just* a human being. He was more like a *phyto-cyborg*—half human mind, half plant mind.

When I got up to the sprawling ranch-style house he had in Esperero, I couldn't figure out where the main door of entry was. There were so many desert and tropical plants around the adobe walls of the house that they obscured the passageway. Then I heard a woman's voice calling me from a pile of boulders twenty feet away from the house.

"Hi, I'm Mahina. You must be Gary. Welcome!"

I could not have imagined at that moment that within five years, I would be co-founding Native Seeds/SEARCH with that woman with a melodic voice who was sun-bathing on a boulder, naked, above a natural swimming pool in the canyon below her.

Then I noted that there was a balding man with a huge beard sun-bathing on another boulder nearby.

"Hi, I'm Gary. You must be Richard," I guessed.

"No, I'm Andrew Weil. Welcome! You'll find Richard in his office just inside the back door. He's going over some new information on the Seri use of plants that was flown up to him by Ike Russell. Go on in. We'll come in after we dry off."

Naked bodies on boulders? Plant data flown up from an Indigenous tribe of fishermen and women from the Gulf of California? It dawned on me that I had just stepped into another world.

Within three months, I too was living in Esperero Canyon, fifty yards behind Richard and Mahina, in a little brick cabin we called the Bella Rita Home for Wayward Boys. Occasionally, fellow Prescott College alums like Kim "Kimo" Cliffton or other desert rats would bring their sleeping bags in to sleep on the floor while working eighteen hour days with Richard.

The first time I went on a field trip into the desert with Richard and Kimo, I felt as though I had been swept up by a desert whirlwind—a *remolino*, or *chachipira*. I was not yet twenty-four years old but somehow felt called to give protracted attention to the Sonoran Desert edge along the Gulf of California. Kimo confided in me that the Seris (Comcaac) called Richard "The Man Who Cuts the Tops Off Plants," because of the intensity with which he collected plant inflorescences for herbarium specimens. Richard had offered both of us a free trip into the Gran Desierto of Sonora adjacent to the Gulf and the Río Colorado, perhaps to see if either of us greenhorns he was bringing along was fieldworthy enough to survive on his projects.

First off, he requested that we help put four gigantic balloon tires he had purchased onto a four-wheel-drive Toyota Land Cruiser that a wealthy woman in Tucson had loaned him. The kind plant doctor offered to cover all the food for the trip, if we liked Dundee orange marmalade, Ry-Krisp crackers, and caviar for breakfast, then nothing else for the rest of the day, except water, until the sun went down.

But then if by chance either of his prospective assistants happened to collect a specimen of plant new to Sonora or to the Gran Desierto flora, he would reward the honoree with a hashish-heavy Alice B. Toklas brownie, courtesy of our Esperero Canyon friends.

We took him up on these generous but eccentric offers. We took off after three days of tire changing and packing on a desert safari into the driest reaches of Mexico. We drove three hours west by southwest from Tucson, as the eccentric botanist rattled off a litany of names of all the plants he saw along the way. He recited nonstop dozens if not hundreds of names of the shrubs and wildflowers he was spotting along the desert roadside in a cadence like Allen Ginsberg reading "Howl."

A half hour before crossing the border, we entered the hyperarid landscape that some of our friends later called The Big Empty. We fishtailed on unpaved roads across sandy flats to reach the base of a giant dune, then parked, leaving the keys on the dashboard. I realized that I had arrived at where I wanted to spend the rest of this ephemeral life.

Gary Nabhan on the dunes adjoining the Sierra del Rosario, March 1975. Richard Felger archive.

For days, we saw no other person, only the distant glimmer of car and truck lights at night on a border highway twenty to thirty miles to the north. We routinely climbed up and down two-hundred-meter-tall sand dunes in dune fields that stretched to the horizon, far more massive than Old Baldy back near Lake Michigan, where I grew up. We worked from just before dawn to just after dusk, encountering a dozen plants new to science in the sandy dunes spilling over the slopes of the Sierra del Rosario. We'd get into camp, and before we pressed each plant between two layers of paper, we'd count stamens, petals, and sepals, feel the stems for stickiness or bristliness, and smell the roots. All senses on deck.

A pivotal moment occurred the morning after the last pot brownie was consumed. The mad scientist had discovered that dermestid beetles had eaten the rest of the Ry-Krisp crackers, and that the jars of caviar and marmalade were completely empty, so he was suddenly ready to break camp.

"Desert expedition called on account of no rain nor food," the droll desert scientist proclaimed.

"Let's head over toward the Colorado River delta. Time to switch to explore the delta and seacoast, where at least we can get some fish to eat."

The only body of water seen for the first eight days was at the desert oasis of Quitobaquito, right on the border. No such springs or ponds in sight for fifty kilometers (thirty miles) in any direction, so we decided to take advantage of it. We skidded our sunburnt bodies under the barbwire fence and skinny dipped with no one else in sight.

"Packed within this tiny oasis are probably more plants and animal species than in the surrounding hundred square kilometers of stinkin' hot desert," the botanist lectured, even while soaking in the refreshing waters of the spring-fed pond.

"It even gets vagrant pelicans and ibises blown off course from the coast. Someone ought to survey its entire biota someday, and its connections through migration to the ocean," the botanist mumbled as he skidded back beneath the border fence.

Duly noted. Didn't know when, didn't know how, but that modest proposal was just too good to forget. I ended up working with him on that biotic survey of Quitobaquito for two decades.

We drove next to the far northwestern corner of Sonora, through low sandy swales to the fishing village known as El Golfo de Santa Clara. We came to a halt at an open-air seafood restaurant overlooking where the Colorado River waters from the north once drained into the Gulf of California in front of the village. After a lunch of fried corvina and Tecate beer with limes, the botanist rummaged through the desultory CONASUPO grocery store for his favorite foods.

Unfortunately, it did not offer caviar, orange marmalade, or Ry-Krisp crackers, so he grabbed some tortillas, canned beans, chiltepin hot sauce, and ice-cold beer. We meandered north on a series of mesas until it was possible to park right above the vast dewatered delta of the third longest river in North America.

We scanned the vast flat expanse below us, which once flowed with three-meter-high tidal bores where the sea occasionally backwashed into the river. We spotted a few truck tracks below that wriggled through the sand that skirted the muddy edge of the Colorado River delta. Descending from the mesa, Richard said "so long" to the northernmost reaches of the Gulf of California, then parked fifty meters above the mud-cacked delta itself.

On the way through the low desert scrub, we found a rarity—a parasitic plant devoid of chlorophyll called *Pilostyles thurberi*. It was growing like little warts on the stems of indigo bushes. Most of its former habitat had already been converted to irrigated fields for growing alfalfa and cotton.

He pointed, and we peered north toward the former camp ironically called Riito for the little that remained of the river. We futilely searched for a little water, but none was to be had. Dried up due to the damming of the Colorado. I glimpsed an expanse of dry mud cracked into pieces like a jigsaw puzzle that was once regarded as a paradise much like the delta of the Tigris and Euphrates, halfway around the world.

It had become an orphaned ecosystem, starved of its essential ingredient: floodwaters. I wondered if there were any extant patches that exemplified what the ecologist Aldo Leopold had once eloquently described as the "green lagoons" of the Colorado River delta. When Aldo and his brother Carl canoed through in 1932, it had been flush with all manner of birds, reptiles, mammals, and fish.

What wildlife was left was in sorry shape. We stumbled upon the rotting carcass of a giant leatherback sea turtle that had been strangled by fishing nets before drifting ashore. It was easily two to three meters long, bigger than the land tortoise I had seen in the Galapagos just two summers before.

Richard was mortified. We stood over the beautiful green and mahogany brown body of the largest and rarest reptile left in the seas on this earth. Too shocked for words.

After minutes of silence, he began walking back to the Land Cruiser.

I heard the Guy Who Cuts the Tops Off Plants repeating a phrase from Cmiique Iitom—the Comcaac (Seri) language—over and over.

Kimo asked, "Richard, what are you saying?"

"*Xiica cmotómanoj*. It is the honorific term the Comcaac give to leatherbacks, when they are in their presence. Instead of the usual name for this sacred being, *moosnipol*, they treat it as a special sentient being, almost like another human. The phrase means something like 'the one who weeps like us.'"

We drove back into the desert in silence, and across its shimmering sands, silent. It took a good amount of time before everyone's eyes were dry again.

I worked off and on with the mad, sometimes sad, always brilliant scientist for a few more years. Then one day, he called me and said simply, "I guess I've taught you all I can teach. It's best you find some other mentors, innovative ones, because I trust you'll do some interesting things on your own from here on out."

My throat was dry.

"Thanks, Richard." I couldn't articulate anything else.

After a few years' break from working with one another, we somehow began working together again. Biodiversity in the Sierra Madre Occidental. Wild relatives of crops found along the Borderlands. Flora and ethnobotany of Quitobaquito. The list kept growing.

One of the last best times I saw Richard in the desert, he had agreed to lead a field trip to the Cañón del Nacapule for NextGen conference attendees in Guaymas, Sonora. Most of the those who hiked with us that morning were a third or a fourth his age. His detailed knowledge of the flora there thrilled every one of us. He was like a superstar finally getting credit for a life well lived in the green and gray world at the desert's edge. It was like seeing Leonard Cohen come off retirement to play his songs with a punk rock band. I realized that Richard had always lived at the edge—there was nothing mainstream about him.

For teaching me how to live a little like that, I remain more grateful for Richard's renegade example than I am for all the equally generous help of any of my fine classroom professors. Not only did he never fit into a box, he never even noticed any boxes because he was eagerly walking past them to see what surprising plants might lie at the desert's edge.

19

# NO ORDINARY MORTAL

Mark A. Dimmitt

I HAVE BEEN HOOKED on plants since early childhood, but all my degrees are in zoology. My friendship with Richard Felger was a huge factor in my pivot back to botany. Many people may view botany as a sedate, monotonous activity that takes place mostly in the quiet halls of herbaria. It can be that, but working with Richard, especially in the field, was intensely energetic and often exhausting.

In the fall of 1967 I started graduate school in the Zoology Department at UCLA. I decorated the desk in my basement office with a butterfly orchid (*Oncidium* [*Psychopsis*] *papilio*), illuminated by a twelve-inch fluorescent lamp. I had been there about two weeks when a strange man walked into my office looking for my office mate, Phil Regal. Without introducing himself, he said, "Oh, that-that-that *Oncidium papilio* can't grow here; it needs more light." I replied that I had been growing it under that light in my home for a few years. Two things happened in that first minute: I was impressed that this guy recognized my rare orchid, and he was impressed that I could grow orchids under such conditions. That was my first encounter with Richard Felger. In the ensuing five decades he profoundly influenced the course of my life.

In the late 1960s Richard was senior curator at the Los Angeles County Museum of Natural History. I visited him there a couple of times, and before I knew it he had roped me into volunteering to help mount and label herbarium specimens. (He was always very skilled at "recruiting" labor.) One day he invited me to his parents' home to see his living plant collections. I discovered two groups of plants that day that changed my life. As we parked in the driveway, the first thing I noticed were strange palmlike leaves towering over the house from the backyard. They turned out to be huge cycads (*Encephalartos*), of which Richard had dozens. I never learned how he acquired them; no *ordinary* mortal had cycads in the 1960s (except sago palms). Judging from their size he must have had them for quite a few years. He also had a substantial collection of *Tillandsia* species. In those days bromeliads were also unknown in horticulture except to a tiny handful of aficionados. (Richard was a charter member of the Bromeliad Society International, founded in Southern California in 1950.) He gave me a few plants, and that triggered my lifelong passion for growing and hybridizing tillandsias.

In the summer of 1968 Richard hired me as his herbarium assistant at the L.A. County Museum. But I didn't curate many specimens at the museum that summer. Instead, we made a six-week expedition up the Pacific coast all the way to the Canadian border. We collected hundreds of specimens, and I learned that a plant press can be stuffed to six feet high. We filled several of them. They couldn't have all fit in our vehicle. He must have shipped some back to L.A., but I don't remember that. The most impressive thing about that trip was that Richard knew almost every plant we encountered. California alone has six thousand species. He wasn't only a desert botanist.

Then I completed my PhD in zoology at UC Riverside and spent four years as a wildlife biologist in the Bureau of Land Management. By 1978 Richard was the research scientist at the Arizona-Sonora Desert Museum in Tucson. A major upheaval there resulted in numerous job vacancies, and Richard urged me to apply for the curator of botany position. Two factors were in my favor. Richard was head of the search committee and knew of my horticultural experience, and all the other applicants were herbarium botanists. The museum wanted someone who could grow live plants, and because of Richard, I got my dream job.

In the past half century I've had many great adventures with Richard. One of the more memorable was in 1985; it illustrated Richard's passion for his profession. On an earlier trip to Sonora, Richard had found a new species of what was then *Macrosiphonia* in the dogbane family, but it wasn't in flower, and you must have flowers to describe a species. He knew it should be in flower during the summer rainy season. Both of us were extremely busy with other projects and time commitments, so we pancaked a trip over a summer weekend. We left Tucson early Saturday morning, drove to the Guaymas area, and collected plants on Cerro las Avispas in the Sierra Baviso and in Cañón del Nacapule all afternoon in weather that made a sauna feel cool. That evening we rented a hotel room and pressed plants until midnight. We got up at 4 a.m. and spent Sunday collecting all day, then returned home at midnight. That was two back-to-back twenty-hour days, a full work week over a weekend. Richard was so busy with a plethora of projects that it took him ten years to get around to describing our collection from that weekend. It was published ten years later as *Telosiphonia nacapulensis*.

We made another discovery on that trip. Cerro las Avispas is in a narrow zone of intergradation between the two forms of the barrel cactus *Ferocactus emoryi*. North of that zone the species has cylindrical stems about a foot in diameter and usually up to three feet tall, with long recurved central spines and deep red flowers. To the south it is a massive barrel, two feet in diameter and up to six feet tall, with short yellow spines and bright yellow flowers. From Las Avispas westward to the Gulf of California coast, the two mix in this narrow band, and we found a full range of intergrades from one type to the other. We even found a chimera that had yellow flowers on one half of the stem and red flowers on the other.

Richard had little respect for mainstream customs and rules; in fact, he relished violating them. For example, he wanted a swimming pool at his rustic old house in the Catalina Mountains foothills. The only level ground on the property was the driveway, so he installed the pool in front of the garage. The realtor was probably apoplectic about what that would do to the salability of the property, but Richard didn't think that way.

Richard struck another blow to convention and homeowner association regulations when he moved into a big corner lot in the historic, wealthy Sam Hughes neighborhood east of the University of Arizona. He replaced the required lawn with a vegetable garden. For several summers the house was nearly hidden behind a big patch of six-foot-tall corn. He grew old standard varieties whose sugar turns to starch within hours of harvest. With at most a day's notice about the crop reaching peak sweetness, he would invite several friends over to gorge on the freshest possible corn on the cob. All of Richard's homes morphed to reflect his quirky personality, and of them, the Sam Hughes house stood out the most prominently.

20

# HUNTING HALOPHYTES AND DESERT PLANTS

James Aronson

I WOULD LOVE to have been with Richard Felger when he joined a delegation of U.S. scientists on an exchange visit to Uzbekistan in the early 1970s, hosted by the Tashkent Botanical Garden and Institute of Botany. And I think of him even now whenever I encounter nomadic or transhumant peoples, or mesquite trees (*Prosopis*) or succulents in Argentine Patagonia, the African Sahel, Australia, and other drylands where I continue to roam. But I did have the good fortune to travel and botanize with Richard in a number of hot deserts during the 1980s. We went hunting xerophytes and halophytes of known or potential economic value and of inherent value to us as scientific naturalists and ethnobotanists. We went to places neither he nor I had ever been, and to a few that only one of us had seen.

Richard and I met in 1981, when I was in my twenties and working in the Negev region of southern Israel. I was hired as a plant hunter and introduction officer at the Hugo Boyko Applied Research Center in Beer Sheva. Hugo Boyko and his wife, Elizabeth, were possibly the first people to consider and actually test the possibilities of agriculture based on halophytes grown with seawater.

Richard was working part time studying, advising on, and collecting new halophytes in association with the Environmental Research Lab in Tucson, where James O'Leary and Ed Glenn were pioneers in the same field as that launched by the Boykos. When we met, Richard was ready to enlarge his search for new arid-land crops and other useful plants in arid and semiarid lands where he had not traveled. I emphasize halophytes because both Richard and I were charged by our institutions to hunt for halophytes of potential use for agriculture, horticulture, and animal husbandry.

Although I was nineteen years younger than Richard, we found each other sufficiently simpatico to envision traveling together to lands unknown. I recall now with pleasure some great moments, and some funny ones, on the road with Richard. Between 1981 and 1986 we traveled to a few chosen places that particularly intrigued us, all megadiverse and fascinating, both botanically and culturally. We visited Israel's Negev and Egypt's Sinai Deserts, and made two trips to South America, one to the hyperarid Atacama Desert of northern coastal Chile and the rain shadow desert of El Monte in Argentina, and the other to the Venezuelan side of the Guajira Peninsula. We also visited Baja California and made a trip to the Thar Desert of Rajasthan in northwestern India, along the Pakistani border. Between trips we shared some relaxation time in Richard's home and garden in Tucson, Arizona, and he came once to mine in southern France, bearing about one hundred cuttings of succulent plants from his collection. He also brought a sapling of the marvelous *Catalpa* x *Chilopsis* hybrid tree, which is still alive and thriving in my garden there some forty years later.

During these trips we shared joy in seeking new crops and candidates for dryland agriculture, as well as candidates for horticulture and industry for North America, the Mediterranean Basin, and all drylands. We also sought new shade trees and ornamentals, multipurpose ground covers, grasses, and fodder shrubs that are inherently tolerant of prolonged or seasonal drought, high temperatures and insolation, desiccating winds, and brackish depauperate soils. Of course, we also prospected plants in favored habitats: oases, natural wetlands, riparian corridors of ephemeral arroyos, wadis, or washes. We especially sought halophytes of possible use, or already in cultivation with drip irrigation, surface runoff, old-fashioned flood irrigation, or no irrigation at all. Also, we pored over unknown land races and varieties of vegetables, fragrant or useful sap bearing, and herbal teas and medicinal roots known to and used by local and Indigenous peoples. Selected varieties of fruit-bearing trees were fair game, like the delicious varieties of grafted jujube (*Ziziphus mauritiania*) cultivated for centuries in Rajasthan, and the dozens of leaf-bearing chenopod shrubs or diverse grasses and sedges appealing and palatable to camels, donkeys, horses, goats, and sheep.

We imagined the taxa and cultivars we saw growing under cultivation in settings where desalinated, brackish, or saline water or even seawater might be used, alone or in mixture with nonsaline, or else with just enough irrigation the first year to help the plants get established and then be left to fend for themselves. Wherever we could, we talked to local botanists, naturalists, pastoralists, farmers, and gardeners. Once, in India, our hosts showed us their trial plots of jojoba and wanted to know what we thought. In Chile, we were shown *Atriplex* trials, and we shared our stories of acacias, *Aeluropus*, *Atriplex*, guayule, and of course *Prosopis* (mesquites).

Richard is famously associated with the group of desert trees in the genus *Prosopis*, and in this context I must mention above all the month-long trip that he and I took to the Thar Desert, in Rajasthan, India, in March 1982.

The Thar is the major locality today for one of the most iconic of all desert trees and one of the most exceptionally interesting from agroforestry, environmental history, and ethnobotanical perspectives. I refer to the khejri, *Prosopis cineraria*. Outside of Rajasthan, the kherji (also known by other vernacular names) occurs in other drylands of India and formerly in Iran, the United Arab Emirates, Saudi Arabia's Empty Quarter, and no doubt, much more widely in southwest Asia. Today it is sadly rare or extinct outside the Thar. This may be a case of "enigmatic" Old World dryland trees whose extant biogeography is puzzling and suggests very early intentional introductions by people that affected the distribution, genetic diversity, and evolutionary ecology of these trees. Only in Rajasthan, to my knowledge, has *P. cineraria* been maintained, cherished as a multipurpose agroforestry tree, and indeed considered sacred.

Khejri is the framework species of an ancient agroforestry system and the stuff of legend. In 1730, the heroic women and men of the Bishnoi villages led, at very high cost, a successful campaign to protect their trees from the woodcutters of the Maharaja of Jodhpur. Inspired by their leader, Amrita Devi, 363 villagers—mostly women, it seems—sacrificed their lives to save their sacred, life-sustaining groves of khejri trees.

The search for new halophytes that could be introduced as crop plants has substantially cooled since Richard and I traveled together. In today's crowded, human-dominated, and heating-up world, the work that Richard and I, and notably Gary Nabhan, were pursuing four decades ago and are pursuing to this day, is more timely and essential now than ever, not least the work on economic halophytes.

As a student of natural history, biogeography, ethnobotany, and ecological restoration, I've had many wonderful, inspiring teachers and mentors, including Peter Raven, Dan Simberloff, Andy Clewell, Zev Naveh, and Al Gentry, just to name five. Richard was one of the most influential, because of the way he fluidly combined botany, natural history, and humanism.

As a travel companion in the 1980s, Richard was at times a bit trying for me. A bit like a thin-skinned, long-necked turtle, an image I choose with love because turtles were Richard's number one spirit or totem animal. At the time I was very young and headstrong indeed. I'm sure Richard found it rough going to travel with me on some days too, but he never mentioned it out loud.

I also quickly learned that Richard truly cared deeply about animals and plants, even scruffy halophytes or spiny desert *Acanthaceae* and nightshades that most people would barely glance at. He felt kinship with nonhuman animals and with plants in a way that is rare today wherever you go.

He also had the same deep respect and empathy for people but especially desert- and semidesert-dwelling peoples, both traditional and Indigenous, with a sense of and identity with place, and real kinship with the plants and animals there. But the trees, landscapes, and peoples of drylands were in a poetic, spiritual sense Richard's "country" and family at the same time. Richard was truly a citizen of the world.

I admired Richard's ability to bridge the worlds of natural science and poetry and his deep sense of beauty both of the natural world and human artwork, including masks and other everyday objects both domestic and ritual. Richard was also a model of patience and generosity, most of the time. His brilliant immersive book with Mary Beck Moser, *People of the Desert and Sea: Ethnobotany of the Seri Indians*, is undoubtedly the most important ethnobotanical record ever made in the New World. He was an independent thinker yet respectful of his friends' and colleagues' opinions and priorities even if he didn't agree with them. Richard was a scientist "in the round," who like Nicholas V. C. Polunin, another great scientist, activist, and mentor to many, deserves to be known and remembered for generations to come.

21

# STOP! PLEASE STOP!

## *A Remarkable Trip to Mulatos*

Alberto Búrquez

I VIVIDLY RECALL the winter of 1969. I was in my last year of secondary school. We were camping at the outlet of Cañón del Nacapule, a remarkable gorge near Bahía de San Carlos in central Sonora. That school outing most certainly marked my first encounter with Richard when an amazing convergence unfolded near the canyon's entrance. It could have been framed into a nostalgic Norman Rockwell illustration—a crowd of teenagers looking at a crazy young gringo with a stack of pressed plants. Overhead, a squadron of B-25 Mitchell bombers roared toward the filming location of *Catch-22*. This chance occurrence was a harbinger of momentous things to come: that crazy young man was Richard Felger.

Twenty-two years elapsed before our formal introduction in Hermosillo during a drylands ecology conference. By then, I had become a botanist and ecologist, had immersed myself in Richard's scholarly work, and wanted to meet the person behind that research. Following the conference, he graciously extended an invitation for me to visit him in Tucson. Our time was marked by extensive discussions, knowledge sharing, generous hospitality, and unlimited access to photocopy scientific papers and chapters from his extensive library—a foundation that solidified a lifelong friendship.

Our bond extended beyond academia. Countless reciprocal visits ensued, marked by field expeditions and extended sessions devoted to identifying pressed plant specimens. Richard became a cherished part of my family. His daily calls to his mother, the gifts he brought for our children, the field excursions, and our long sessions of sorting, pressing, and labeling plant specimens were hallmarks of his visits.

In 1995, after countless desert treks and botanical collection ventures, I extended an invitation to Richard for an immersive journey into the heart of the Sierra Madre Occidental. I knew he was eager to reach far into the Sonoran highlands. My plan was to compile a floristic list and document the vegetation of Mulatos, an old mining enclave nestled close to the Sonora-Chihuahua border. Richard enthusiastically accepted the invitation, urging me to gather as many newspapers and plant press cardboards as possible. Despite my frantic efforts to collect cardboard boxes and newspapers, Richard took one look at my big stacks and announced that we needed twice that amount! I was a relative beginner, and he was a big leaguer.

So an extra day was dedicated to procuring not just two additional plant presses with the necessary paper and cardboard, but also an array of specific items: robust plastic bags, parachute cord, corn and wheat tortillas, assorted nuts, and raisins.

We left Hermosillo before dawn and set out for Sahuaripa, a historical town that signaled the character of the real Sierra Madre Occidental. Our full day's journey amounted to 220 kilometers and should have taken us only four hours on the road, but merely half an hour from Hermosillo, Richard yelled, "Stop, stop!" The western slope of the pass at El Orégano boasted a magnificent saguaro grove, transitioning from pristine thornscrub to the characteristic Plains of Sonora. It was October, with hints of red and yellow foliage adorning the hillsides, courtesy of burseras, crotons, and tropical ocotillos. Richard jumped out of the car and started filling a large plastic bag with cuttings. What was promised as a brief pause turned into an hour-long collection excursion. It ended only at my insistence.

We continued to the town of Mazatán, whose cheese is famous throughout Sonora. The locals are so proud of it that they made a monument to the cheese: a cowboy, his wife, and two kids stand around a big cheese wheel at the town's entrance. At Richard's directive, we bought cheese and paused for breakfast, less than one-third of the distance to our destination. This slow pace set the rhythm for the remainder of our trip.

North of Mazatán, the Sierra Huérfana greeted us. Richard gave me a long explanation about the botanical treasures that the isolated range harbored and the importance of sky islands to species creation and differentiation. The top of the Sierra is a heart-shaped mesa carpeted with oak woodlands. Just below, the escarpments are covered with tropical deciduous vegetation surrounding the mountain and giving way to the desert and thornscrub beneath. This range, the southernmost metaphoric core complex of Tertiary mountains in the Southwest, stands alongside prominent counterparts in Sonora (Magdalena, Aconchi, and Pozo Verde) and Arizona (Santa Catalina, Rincon, and Tortolita, among others). All of them were forged during the early Basin and Range extension processes. The extensive dry forests on Sierra Huérfana mark the northwestern edge of highly diverse tropical vegetation. Nevertheless, a few tropical trees—reduced to small shrubs that periodically freeze—find sanctuary in protected canyons of the Santa Catalina, Chiricahuas, and other mountains in Arizona. Richard gave me a list of plants I had to look for up on the mountain so that we could collect and deposit them at the university herbarium for him to examine later.

Numerous stops dotted our journey, as Richard's insistent requests echoed: "Alberto, stop please, please stop! Backtrack a hundred meters; I think we have stumbled upon a new species distribution record!" Amid these pauses, we started our descent to the Río Yaqui valley, downstream from the massive El Novillo dam. In a mere 4 kilometers we plunged 650 meters down to the river-crossing bridge. The main and secondary canyon slopes are draped with lush tropical vegetation—fig trees, climbers, exotic hardwood plants, and various species of elephant trees enveloping the landscape. Midway, we found ourselves collecting plants while precariously perched on the road's narrow shoulder and dodging loaded cattle trucks. Richard's joy emanated not from grandiose trees or vibrant flowers but from tiny inconspicuous plants that gave us coveted northern distribution records of remote tropical taxa.

La Estrella, a village by the Río Yaqui, marked a natural transition between the arid Sonoran Desert to the west and the Sierra Madre Occidental to the east. Our ascent rose toward the sierran towns of Bacanora and Sahuaripa. Despite Richard's pleas, I resisted further stops. He wanted to collect *Dioon sonorense*, an ancient cycad lineage diversified in Mexico. This species, now designated as *Dioon vovidesii*, has been recognized as distinct from its southern counterparts of Álamos and northern Sinaloa. Near the renowned sierran town of Bacanora, famed for lending its name to the traditional Sonoran agave beverage sourced from northern populations of *Agave angustifolia*, Richard again begged to look for the coveted cycad, but I turned a deaf ear to his pleas for another brief detour to secure a specimen of it.

Our route led us to an ancient settlement where Sisibutari, the Ópata leader in the early 1600s, beseeched Jesuit missionaries to bless the town as Sahuaripa, signifying "place of yellow ants." There we veered from Highway 20 onto tortuous sierran back roads and pushed on toward Mulatos. What was typically a four-hour trip to Arivechi stretched into a seven-hour adventure due to our frequent stops for collecting and pressing plant specimens.

Morning turned to afternoon and marked the second leg of our journey—a lengthy dirt trail requiring four-wheel drive on certain sections. Moreover, Richard's stomach rumbled with hunger, and soon I learned how difficult it could be to deal with a hungry Richard.

Arivechi sits at the center of a long river valley with a verdant agricultural belt. Riverbank farming enabled the Ópata people to practice highly productive irrigation agriculture. The same crops that the Ópata developed are grown in the fields: maize, beans, squash, chilies, and cotton, some of the primary native crops, and alfalfa, garlic, and wheat, some of the Old World crops brought by the Jesuits. Richard's familiar pleas grew louder. "*Stop, please*." He had a vast interest in crops, particularly crops that could be grown with little or no water. Many times, he had told me that dryland crops were the key to human sustainability and long-term survival, and of course he was right!

As we headed into the Sierra Madre Occidental, our conversation changed from the wonders of biological diversity to the lawlessness of the deep Sierra, the fields of *Cannabis* and opium poppies, and the deadly ruthless people hiding in remote places. I shared that the Ópatan etymology of Arivechi was "place of skulls," but that only worsened Richard's mood.

The road rapidly gained elevation, and the vegetation transformed into oak woodlands interspersed with thornscrub. A pronounced ecological slope effect was evident on the hillsides, prompting Richard's desire to collect specimens from north- and south-facing slopes. We discussed how this ecological variance arises from contrasting solar radiation exposure. In the Northern Hemisphere, north-facing slopes have a cooler, umbrous, and more humid microclimate, while south-facing slopes are drier and exposed.

Richard grew increasingly contentious about my reluctance to make stops. And I grew apprehensive about reaching Mulatos before nightfall, particularly so after our conversation about drug cultivators and the ominous connotation of Arivechi's name. But I relented, and we briefly stopped for a meal alongside the road. The spot was scenic, with the most gorgeous view of bluish range after range of mountains. While I prepared sandwiches, Richard went on a collection frenzy, and then we ate and had beverages, apples, and oranges. Finishing *le petit repas*, Richard's glucose rose, making everyone happier. In the company of friends, even the simplest meal becomes a banquet, eases anxiety, and energizes plant collecting.

As we ventured beyond the town of Tarachi, our journey took a delightful turn, guiding us into higher-elevation pine and oak forests. A stunning metallic blue-green tarantula gracefully crossed our path, adding an unexpected touch of nature's beauty to our experience. As dusk approached and we reached the quaint village of Matarachi, the waning sunlight cast long shadows of the pine trees, creating an intricate tapestry on the landscape. But we grew increasingly apprehensive about getting caught in the dark. We were driving my university's diminutive red Suzuki—dubbed El Chiltepín, the little red pepper. We continued along increasingly narrow roads toward La Muela pass, and as the road became more difficult, we were grateful to have a four-wheel-drive car. We relied solely on a folded map for navigation. After a particularly slippery ascent, we paused to reorient ourselves, anxious to reach the old mining town. We parked to check the map again in a lovely meadow surrounded by a towering pine forest. An eerie silence enveloped us, pushing us to drive on.

Nightfall greeted us as we arrived in Mulatos. What a relief! The mine's mess hall awaited our arrival, and its cook, a woman from Guerrero, welcomed us warmly. She had a beautiful smile, typical of the Costa Chica people. Her culinary offerings were great. The subsequent days were a flurry of intense work, exhausting hikes, and endless specimen pressing. We reveled in exploring this largely uncharted territory, compiling an extensive species list, and crafting a map of the region's vegetation. We ventured by mistake into a *Cannabis* field and ended up having coffee with its keepers. Richard excitingly stumbled on a new species of euphorb, which he and Victor Steinmann named after me: *Acalypha burquezii*—a flattering honor! After a week, and laden with botanical treasures, we departed for Hermosillo, where Richard diligently curated the specimens and integrated much of our deep sierran venture into his book *Trees of Sonora, Mexico*.

Since our expedition back in the nineties, immense gold mines in the area have expanded in both scale and number, but the people of the region are learning about the importance of preserving ecosystems, key species, and biodiversity.

And since this adventure, Richard and I have collected plants at many other localities in Sonora, including his beloved islands of the Gulf of California, the tropical forests of eastern and southern Sonora, and the coastal strip from the Colorado River delta to the thick mangroves of northern Sinaloa. We had many other adventures over the years, but since the first trip to Mulatos, I've always stopped the car when asked to collect plants. "Pressing matters," he used to say. I also made sure I had enough food for my dear friend.

Richard's maxims live in my heart: "Pax et Prosopis. Pax et precipitation. Pax et Pachycereus. Saludos y Sideroxylon. Lightning and thunder. Mucho rain . . . y ahorita viene mas. Happy May Day. Workers unite in syncarpy. Mucho many thank you. Happy monsoon. Mucho gracias. Pronto, prontito. Right now, not Monday." And "Stop. Please stop!"

# 22

# RICARDITO

Angelina Martínez-Yrízar

RICARDITO, AS I called him, was a one-of-a-kind man. Since the first time we met back in the early 1990s in Hermosillo, I realized he was extraordinary and grew to appreciate him even more over the years. Our long-lasting conversations at home covered plant names, lists of Sonoran Desert and tropical deciduous forest species, and the many places in Sonora he wanted to revisit again and again to collect data or to find new species for his monumental botany books. His love for Sonora was inspiring—he would enthusiastically describe the secrets of the flora of the Gulf of California islands, the Sierra Libre, and the Cañón del Nacapule, which he knew better than anyone.

He was incredibly generous to my husband, Alberto Búrquez, and me. On our first visit to his home in Tucson, he invited us to stay, browse his extensive library, and freely photocopy as many pages of books and reprints as we wished to bring back to our lab at the Universidad Nacional Autónoma de México in Hermosillo. During those years, our only way of getting papers, books, or scientific literature was by searching the University of Arizona Science Library or by requesting them by "snail mail" from the authors. At any time and without hesitation, he shared his unpublished information with us on species distributions and local flora, to a degree that on one occasion he gave Alberto an unpublished plant list with a serious warning to "keep this out of enemy hands."

He spent hours on end in our lab identifying plant specimens that we had collected from our field sites near Hermosillo, though not entirely out of generosity, since he was clearly eager to see what we had found. He frequently looked at our plant presses and complained that "these ecologists do not know how to properly press plants!" Then he gave us a lecture on curatorial matters on the spot. His passion and respect for plants were immense. When introducing a friend or colleague to him, he once smiled and told me, "I may not recall his name later because if I keep it in my memory, I will not be able to store a plant's name." He was so knowledgeable about plant species that I could email him a photocopy of any Sonoran Desert specimen, and he would have no trouble identifying it.

Many times, he stayed at our home in Hermosillo, bringing presents to our then-little boys Bertie and Emilio, as well as seasonal delicacies for us. On one occasion he brought us a special gift: a bunch of carefully wrapped potted plants, among them two small *Aloe* trees and two *Welwitschias*. After more than two decades growing in our backyard, the aloes are almost four meters tall each, and although we lost one of the *Welwitschia* plants one year, the other continues growing its single pair of long-lived evergreen leaves and producing cones every year. So we have a constant reminder of Richard in our garden. As soon as he arrived at our house and before anything else, he would ask us to use our telephone to call Silke and his mother to let them know he had arrived safely in Mexico. He did this every time he traveled, and I felt moved by that. Dinner conversation was always a learning experience and fun. Early the following morning, getting ready for his trip with Alberto, he never worried about essential field supplies like food or water, but he fussed about not having enough newspapers for his and Alberto's plant presses. When coming back after a long and exhausting field trip, he would not rest without first labeling and properly pressing each of the dozens of specimens they had collected.

Ricardito was so intent on his work that after sending an email to Alberto requesting either some information or a new figure for his next book, he would hardly wait for a reply before asking me to please tell Alberto to reply "Right now, not Monday!" Going to the field with him was an adventure, but at the same time, we could only advance at a tortoise pace (*¡a paso de tortuga!*) because he went inch by inch making sure he did not miss a single plant. Attempting to explore a small hill the first day, we finished only the foothills and had to leave the hike to the top for the next day.

He was tireless. I am amazed that he was working intently on his *Flora of Guaymas* manuscript only a couple of months before he died. Earlier he had asked Alberto to write the preface, urging him to have it ready as soon as possible. Now I understand why Ricardito was in a hurry to finish it, even as he underwent medical treatment. He always ended his emails to me with "MUCHO thank you—pax et prosopis, R." Peace and mesquite. *¡Gracias Ricardito por tanto!*

23

# MAKING PLANTS SPEAK

Humberto Romero Morales

GETTING TO KNOW the book *People of the Desert and Sea: Ethnobotany of the Seri Indians* was a great advance for me.[1] Information from this book by *el doctor* Richard Felger connected me to all the work that had been done before. I didn't know this information very well as a child, but when I grew up, there it was. The names of the plants and the people collecting them caught my attention.

I first met Richard when I was playing a game with a wheel toy cut out of the stem of *Stenocereus thurberi* [organ pipe cactus], pushing it with a forked stick. Becky Moser took a photo of me, and it is in the ethnobotany book [page 164, figure 13.6; Humberto is the child in the middle]. I especially got interested when I took a plant to him when I was about the same age, six years old. He gave me a bag of chocolate for having brought it. Later, I brought him another plant, *caal oohit* [*Triteleiopsis palmeri*, blue sand lily], but this lily had an [unusually] big flower. I collected it in a place with white sand dunes where

1. This text is transcribed, translated, edited, and annotated for clarity from a conversation in Spanish between Humberto Romero Morales and Ben Wilder on October 7, 2023, at Humberto's Xaaslca Án Botanic Garden, just north of the Comcaac village of Punta Chueca, Sonora.

they grow south of Desemboque. It was really big, and it looked like an onion in the center, with a central stem that comes out above the ground. He was really interested, and I could see he valued what I brought him. He gave me a few pesos, and I went away content. That recognition and motivation have stayed with me since childhood.

That someone valued a plant in that way gave me a lot of interest. I wondered, what are these books written in Cmiique Iitom [the Seri language] or Cocsar Iitom [Spanish] or Maricaana Iitom [English]? I wanted to learn from him or from Becky Moser and Eduardo Moser, the writers. I started from scratch. My people and relatives are very intelligent, but why didn't I know anything? I decided then to focus on the vegetation; that is something I could do. I focused on that when I worked on Isla Tiburón as a guide assisting in the hunts of *mojet* [bighorn sheep]. I took care to study and learn the vegetation. When scanning with binoculars wherever we were, I was gathering information about the plants, what the vegetation was there. Based on direct

Richard and Humberto pressing plants and recording data after a day of collecting, with Ernesto Molina looking on, Tecomate, Isla Tiburón, September 2007. Photo by Ben Wilder.

observations with a spotting scope, I knew what plants they were eating. I noted at what time and how long they would eat which plants. I pulled together this information on the diet of the desert bighorn sheep, and I presented it in 2000 here at a meeting in the Julio Hotel in Bahía de Kino. It was full of biologists, and I wasn't a biologist like them, but they gave me the chance to publish the data I gathered.

Thanks to knowing Doctor Richard, I learned all about the vegetation that I did not know. I think of him as an elder who taught me things, like an old Cmiique [Seri person]. That's how I see him, every time I think of him. I felt a love, an appreciation that you can't imagine.

This was the motivation that Moosni Ilít [Ed Moser] and Rebeca [Becky Moser] gave me, and Ricardo too. I am grateful for the pieces of the puzzle they gave me about plants from various conversations, their publications, and comments. What I learned from him, what he gave me. This teaching, this learning that I have now is because of him. I have joined them, those who study, and I am thankful and happy for that.

At first my knowledge was limited to the few plants in Cmiique Iitom. Then with the ethnobotany book, I added the Latin names. Because of this knowledge, so much more information was indirectly transmitted to me, without anyone telling me anything.

My first copy of the book [the Seri ethnobotany] got so dirty from where I put my fingers as I flipped through the pages I was studying, over and over, that I had to keep getting new copies. These books should be kept in a museum or library. I took notes, taught myself, pronounced, and spoke by myself. This plant is named this; this is called in this way. I spent years practicing by myself.

I asked the elders in my community, but they didn't tell me anything. Which is which? They still didn't tell me anything. Since that first trip I took to the island as part of the bighorn hunts, I have asked about the strange plants there. I asked about a *Randia* because it had flowers. "What plant is this?" I said. Nobody told me anything. I cut a little sample from a canyon on the island and brought it back and showed it to a person. He told me that it's *haalp* [*Randia thurberi*], and it has black pulp inside of its fruit.

That is how my knowledge advanced. I knew I was learning something. I was practicing from the book, teaching myself when everyone else was asleep. I would get up and read: what is the important vegetation? What are the different plant names, and how are they called? With that, I advanced. I came to know *comitin* [*Olneya tesota*, ironwood], *xoop* [*Bursera microphylla*, elephant tree], *haaxat* [*Larrea divaricata* subsp. *tridentata*, creosote bush]. I learned very quickly. At first, I relied on the book, and then later on with what Ben and Richard taught me. I would ask about specific plants, like *Croton magdalenae*: why is this a *Croton*? It is nothing like *moosni iti hateepx* [*Croton californicus*], which is also a *Croton*. It is *hehe ziix capete*. [The small shrub *Croton californicus* is found on beach dunes. The Seri name means "what sea turtle meat rests on." It is markedly and almost unrecognizably different from *Croton magdalenae*, a large fuzzy-leaved plant of the interior canyons of Isla Tiburón.]

From there I made very quick progress with what I had already learned in Cmiique Iitom; I just added the common name and the Latin name. Now there are many changes to the scientific names, so it's a bit tricky, like I need to relearn them all over again. It's not complicated. It is a matter of writing it and teaching oneself. [Humberto once said to Ben, "We Comcaac are going to change our names each year, and then we'll see how *you* manage!"]

Whenever I have a question about something, I always remember Richard, like he is here with me. How quickly he would always reply to me. I miss him. I miss him for how generous with his knowledge he was. If you ask him something, how fast he would tell me. There are many times when I miss him. When I'm explaining things to people, I remember him. He was so connected to the life around us. Whatever plant he would name, he described it, what it is, what family it is. His memory is something historic for me, a great person, unforgettable.

Once I found a *Tragia jonesii*, a beautiful plant, green, shiny. I mentioned to Ricardo that it hurt me, *que me chingo*. I asked, "What is the name of this plant?" He said, "*El que te chingó*," joking with me. A little bit later he seriously told me it's called *Tragia jonesii* [*Hehe cocozxlim* in Cmiique Iitom, which means "plant that causes rash"]. Now I doubt that many of the young people know that there is a *Tragia* like this.

He told me many of the plants that I didn't know in Cmiique Iitom. Like *Prosopis glandulosa* [mesquite], he told me, did you know it has four names in Cmiique Iitom? It's *hoohopam*, it's *heep*, it's *xpeemoja*, and it's *haas*, the current name. I had not heard these other names before.

"You are one of the last *ancestros*, knowledge keepers," I told him. "By your knowledge, you are. You have different skin, but I see you are the same as us. *Gente anciano* like us. I value you," I told him. I remember a lot about the *antepasados* like both of my grandfathers; I remember them, I lived with them, but I did not learn as much as what Ricardo shared with me. "As I am picturing them, I remember them, and I will remember you," I told him. "Okay. Well, thank you very much," he said.

He also told me about *Suaeda nigra* [*hatajipol*, bush seepweed] and *Atriplex polycarpa* [*hatajixp*, *hatajísijc*, desert saltbush], and the other one called *hatajeen* [*Plantago ovata*, woolly plantain]. "We need to change the names," he said. Because they refer to the noble part of a woman, he explained. *Hatajeen*, similar to *hatajix*, which is the body part; *hatajipol*, which is *hatajix cooxp*, white. *Atriplex polycarpa*, salt bush, is edible, and in Australia they eat it; they add it to food.

It is information like this that Ricardo shared, very interesting things that he mentioned. I value a lot when he spoke. Ricardo spoke very few times, but the words he said were real and true. When I talked to him about *Jatropha*, he asked me, "How do you know these things?" It is the information shared by the biologists I'm with, I told him. It's their knowledge that was passed on by them. That's why when someone speaks like that, I listen well and hear them well, and take seriously the things they talk about. That's how I learned. "That's how you can learn, too," I told Ricardo.

How beautiful to know things about other lifeforms. It is an intelligence. An intelligence that giants hold. They are people who make vegetation speak even though they don't speak, but they can speak for them. To me Richard is a giant. A great person. A great mind. A great knowledge.

The Comcaac people have also learned from him. Ben—you and I and others have learned together. I see you as a brother, I see us as sons, I see him as a father.

# 24

# DO YOU WRITE TO BE READ?

Exequiel Ezcurra

I MET RICHARD FELGER for the first time in 1981, when I was starting my graduate research in the Gran Desierto. Crossing the border from the Sierra Pinacate through the Sonoita/Lukeville pass, I used to visit Tucson to check publications in the University of Arizona Library and specimens at the herbarium. Charles Mason, the herbarium director at the time, introduced us, and soon after, Richard invited me to visit him at his home in Tucson. I was awestruck; for anyone interested in desert ecology, Richard was already a legendary figure. A few years before, he had published two extraordinary papers in the journal *Science* based on his work with the Seri (Comcaac) people and his knowledge of the Midriff region of the Gulf of California. One paper described how the Comcaac made a highly nutritious flour out of the seeds of eelgrass (*Zostera marina*), and the second one described the winter dormancy of sea turtles in the Canal del Infiernillo and their sustainable use by the Seris. The papers contained all the novelty and excitement that a thirty-year-old ecologist like me at that time might ever dream of: exploration and the thrill of discovery, the remoteness and the challenges of the deep Sonoran Desert, rigorous science, the wonderful human connection of ethnobiology, and, at the end, being published in *Science*, the most important scientific journal in the world! And as if that were not enough, the first paper made it to the journal cover, which contained a photograph of a Seri doll made of dry eelgrass straw and cloth. I remember it to this day; the cover caption read: "Eelgrass doll made by Ramona Casanova, El Desemboque, Sonora Mexico." That combination of good field biology, a simultaneous interest in plants and human cultures, sound natural history, and rigorous thinking was irresistible. I was, I must confess, dazzled by the knowledge, the intelligence, and the character of Richard Felger. To my surprise, he was also extremely approachable. We became good friends.

In 1981, while working in the Gran Desierto, I encountered for the first time the legendary *pozos*, the mysterious freshwater springs of the salt flats around Bahía de Adair, which had been described two centuries before by Father Kino in his narrative of *Favores Celestiales*, and had been revisited a century later by Carl Lumholtz, who described them in his book *New Trails in Mexico*. I was fascinated by the legendary desert springs and started imagining hypotheses about their origin, their hydrological dynamics, and their associated flora. During a visit to Tucson, I shared my findings with Richard, who became extremely excited by the existence of the freshwater springs in the hyperarid Gran Desierto salt flats. Having read Godfrey Sykes's book *The Colorado Delta*, Richard was quick to realize that some of the aquatic and riverbank vegetation that had been lost when the Hoover Dam was built in 1935, drying up the downstream section of the Colorado River, could be surviving in the flooded areas around the larger *pozo* wetlands. Together, we started a series of trips to the springs of Bahía Adair, collecting plants, water samples, and sediment cores from the salt flats. The exploration of the pozos culminated with a joint expedition in January 1984 with Richard; Ann Russell, a hydrologist from the University of Washington; and Miguel Equihua, from the Instituto de Ecología in Mexico City. Under Anne's leadership, we drilled the clayish mudflats and installed piezometers to monitor the Gran Desierto aquifer while Richard feverishly roamed the springs looking for new plant species. In the *pozos*, I started to appreciate Richard's uncanny ability for natural history exploration and interpretation. He showed me that the middens of sand and clay that surround the *pozos* are actually large accumulations of shells, mostly of murex snails and large clams, gathered in the distant past by early inhabitants of the Americas and accumulated near the *pozos*, where the gatherers used to congregate for water. The shells, he taught me, clearly showed percussion marks, indicating that they had been opened by humans to eat the mollusk inside. The strata in the middens clearly showed the cultural evolution of humans in the desert; while the lower layers contained only shells and a few chert flakes, the upper layers often contained fragments of clay pottery and obsidian. One day we found a huge basalt metate, in all likelihood brought with great pains across the sand dunes from the distant Pinacate volcanic shield. Richard pranced and danced with joy at our discovery: the early inhabitants of the Sonoran Desert not only were surviving on shellfish, but were clearly also able to gather seeds form the halophytes in the salt flats! It was one of the most inspirational and rich expeditions I have done in my life.

In 1983 Richard came to Mexico, where we traveled the Tehuacán Valley. We visited the chinampa farmers in Tláhuac; we went to the Huastec cloud forests of Cuetzalan and explored with exhilaration and enthusiasm every imaginable market from Oaxaca to Zacatecas. With Richard you don't visit local cultures, you *eat* local cultures in the form of pitayas, *chapulines*, tortillas made in a thousand ways, and myriad forms of those most wonderful greens, the Mexican *quelites*, which we bought and stewed, steamed, broiled, or fried in clay comales, and finally ate in the form of tacos and quesadillas made of blue, white, yellow, or purple handmade tortillas, or every imaginable form of gorditas, *sopes*, *tlacoyos*, and *huaraches*. With Richard, cultural and biological diversity were not quantitative abstractions that demand abstruse indices to measure; they are very concrete realities that come to the market in burlap bags and enter into your system through your mouth, your nostrils, and your eyes. For Richard, you measure biodiversity in the richness of flavors and scents. Or, as Frida Kahlo once wrote about Mexico, "*¡Los colores! . . . ¡los olores! . . . ¡los sabores!*"

In July 1984, a few months after our wonderful expedition to the Bahía Adair springs, I was in Bangor, Wales, writing up my doctoral dissertation, and Richard came to a meeting of botanists at Kew Gardens in Richmond, a London suburb. I took the train to London and met Richard at Kew, where he introduced me to an exciting and brilliant party of botanists from Israel and France, all of whom had an acute interest in deserts. The group included Avi Schmida, Dov Pasternak, and James Aronson, some of the sharpest minds in the world to study desert ecology. Discussing the natural history of Old World and New World deserts with this group was a celebration of intelligence and, for me, one of the most enriching experiences of my life, ranging from the vibrant evolutionary questions from a youthful James Aronson to the quiet and deep wisdom of the older and very sage Dov Pasternak, to Richard and Avi's constant intellectual dueling about the uniqueness of desert floras and the extraordinary adaptations of desert plants. I returned to Bangor with a whole basketful of ideas for my dissertation and my future research.

A week later, Richard sent me a message that he would like to visit me in Bangor, and a couple of days later he arrived. It was so extraordinary to host in North Wales my friend and colleague from the Sonoran Desert. We spent time bicycling, eating pub food—fish and chips and British pastry—and drinking gargantuan amounts of locally brewed beers. There, we finished the first draft of our paper on the springs of the Gran Desierto and dreamed together about a large string of protected areas from Bahía Adair in Sonora all the way to the Saguaro National Park in Arizona.

Many decades have passed since, and we nurtured our friendship throughout all those years. Many of our Bangor dreams have come true thanks to the devoted work of myriad conservationists from Mexico and the United States. The Pinacate and Gran Desierto are now a Biosphere Reserve and were recently declared a World Heritage Site (a fact that proves, by the way, that bicycling, British cuisine, and craft ales do help to make lofty dreams come true). Richard moved on into new and always innovative ventures. He stopped publishing whiz-kid papers in *Science* and, rather, started using his time to put his knowledge into books that would actually leave a deep and meaningful Sonoran Desert legacy. Magnificent, superb books like *People of the Desert and Sea*, written with Mary Beck Moser; *Trees of Sonora, Mexico*, written with Matthew Johnson and Michael Wilson; *Flora of the Gran Desierto and Río Colorado of Northwestern Mexico*; and *Dry Borders: Great Natural Reserves of the Sonoran Desert*, edited with Bill Broyles, have put into perspective a lifetime's wealth of information on the Sonoran Desert and its wonderful people.

All the Sonoran Desert researchers and, I dare say, all the country of Mexico owe Richard an immense debt of gratitude for his monumental contributions to expanding our collective knowledge in ethnobotany and the natural history of desert plants. But apart from his scientific legacy, it was his ability to change, to evolve intellectually, and to innovate that never ceased to amaze me. I always learned new and astonishing things at his side, and he always made me think about the true meaning of being a natural scientist. About ten years before his death, we had a conversation about the significance of science and research. I mentioned something about the need to publish some of my own research in journals with high citation factors. Silly me; he gave me one of his shrewd, incisive looks and said:

"Do you write to be cited, or do you write to be *read*?"

In a short sentence, he was able to condense the deep meaning of research and discovery. He never ceased to amaze me.

25

# DARK HORSES, LITTLE TURTLES, AND FEARSOME WOLVES

Gregory McNamee

CUE A SCENE. It's deep night in a tiny town, the streets illuminated by torches. A shadow falls on a wall: a ferocious wolf, to all appearances, sending the townspeople into headlong flight into their homes, where they hide, quivering. Just then the sun begins to rise, and there, in an abandoned butcher's shop, enjoying chain of hot dogs as a snack in good Looney Tunes fashion, is a bloodhound pup.

I am belaboring a dumb metaphor, but to a purpose. I first met Richard Felger along about 1981, when he would come daily, it seemed, to the offices of the University of Arizona Press to go over the editing of a massive manuscript, hundreds and hundreds of pages long, that would later become his book, with Mary Beck Moser, *People of the Desert and Sea: Ethnobotany of the Seri Indians*. I was just starting off in publishing as a proofreader, and my interactions with Richard were at first a simple smiling exchange of greetings, even as other editors averted their eyes or went the other direction, fearing the lupine wrath that they'd been told was sure to come. Indeed, the editor who worked with him on the book lived in abject terror of those meetings, for, ever convinced of the righteousness of her every comma deletion—and the meetings always started with half a dozen sharpened number two pencils—she met with quiet resistance that mounted to occasional wolfish snaps whenever she was wrong, as she all too often was. Moreover, she agonized over lexical choices more than James Joyce ever did, looking up the spelling of every word, consulting the style manuals sentence by sentence, as if to stretch out the editing to her retirement years long into the future and meanwhile guarantee lifetime employment. An office or so away, I could hear Richard, frustrated, and the editor, flustered, go at it, and I marveled at the waste of spirit.

Months went by, the editor went away for a month or so, and it fell to me to pick up the work. Richard and I sat at a table the size of an aircraft carrier deck, dozens of pages laid out for review. He said, "We've been arguing over whether paloverde should be palo verde, one word or two." "Hmmm," I said. "How do you want to spell it?" "One word." "Will it cause a scandal in the profession or bring down a negative review?" "No." "Let's go with it." We finished a couple hundred or so pages of the book quickly and amicably, and from then on I was an ally, having passed a test of some sort.

Then the editor returned, and more months went by, and the book was finally finished. Richard would later recall: "Sometimes I had to fight to retain the most important images and concepts. I seldom gave in and am glad that I held to my convictions. But there is one thing that I regret giving in to: Just as the book was going to press, I let the editors remove a carefully worded paragraph in the preface that said that this book is only a fragment of the traditional knowledge, and we encourage further work to record and preserve traditional Seri knowledge. But the editors said that our book was a definitive work and wanted the paragraph removed."

That was not a decision I had any part in or knew about at the time, and it seems inexplicable to me—knowledge is provisional, after all, and Richard's modest paragraph got right to that point. I'm also a touch surprised that Richard consented to it, for anyone who ever knew him knows that Richard could be, well, difficult at times and dig in his heels as needed. That difficulty, I think, came from an assumption that everyone knew as much as he did and worked as hard as he did, but altogether too few matched him on either score. When we contended, rarely, over a style or a word or a subheading or what have you, Richard would withdraw to study the manual and would almost always return with a rejoinder that brooked no further argument. He cared about such things, paid attention to language as closely as he did to taxa and morphologies, learning and teaching at the same time.

Richard also objected, with many growls, to the price of the book, $65, which amounts to about $195 in 2026 dollars. I was by that time editor-in-chief and *did* participate in making such decisions, but my explanation of how the arcane voodoo of book pricing works, number of pages translating into number of dimes and dollars, never satisfied him. More growls ensued, but in time the book sold its run and paid for itself, and it remains in print nearly forty years later—and now at an entirely reasonable $45 in paperback.

Years went by, with Richard always at work on five big projects at a time. I left the Press to pursue my own work as a writer, and more years went by, and some of those big projects of Richard's turned into big books, among them *Flora of the Gran Desierto and Río Colorado of Northwestern Mexico* and *Trees of Sonora, Mexico*. I would see him at conferences and *pachangas* (informal gatherings) from time to time, hear tell of some blood feud brewing with another scientist over something or another, learn of other projects and collaborations that would bear significant fruit in the decades to come.

Then one day Richard called and suggested we go out to lunch. I was a touch surprised, since we weren't really lunch buddies, but we wound up spending the better part of an afternoon visiting at a place that seemed just right for it: a Tucson sushi restaurant run by Sonoran immigrants of Japanese descent, with chiles far more fiery than wasabi lurking under the hamachi, a perfect meeting of many worlds. Richard, it turns out, was saying goodbye, at least of a kind: he was moving, he said, to the highlands of southwestern New Mexico, where he intended to lock himself into a home office, get his archives together, and set out on the next batch of books that were screaming to be written.

It didn't work like that, of course. Richard labored diligently at all those things, but the Gila Wilderness in his backyard called just as insistently as his beloved Sonoran Desert did, and new projects piled on top of old ones even as he became ever more interested in the problem of how people in our dry country were going to feed themselves as the temperature climbed and the water evaporated.

The years went by, with an occasional email or phone call. Then, somewhere along about 2018, my young friend Robert Anthony Villa—who will one day be enshrined in the same desert rat hall of fame as Richard, I'm sure—came by and told me that he'd been at botanist Mark Dimmitt's house, poking around in the library, and had come across a book of poems that Richard had written in the early 1970s. "Poems? Felger?" I asked. Yes, poems, Felger: For it turned out, as I might have known, that Richard had shapeshifter skills and could move easily among numerous camps and be welcomed by them all, whether a Comcaac hamlet along the Gulf of California or an august assembly of scientists or a finger-snapping Beat Generation coffeehouse in San Francisco, where, way back when, Richard had spent time hanging out and talking shop with Gary Snyder, Michael McClure, Lawrence Ferlinghetti, and Allen Ginsberg.

His little book, called *Dark Horses and Little Turtles*, had come out in 1974, published in a tiny edition run off on a mimeograph machine by poet and artist John Brandi. Now that the book had resurfaced, Robert said, Richard was interested in finding someone to republish it. I thought for a moment and said, "Well, I'll do it," by which I meant that I would reissue the book through Polytropos Press, the infinitesimally tiny publishing house that I'd set up mostly to publish digital editions of my own out-of-print work, but that had quietly morphed into a small literary press with a few other books on the menu.

Robert conveyed that to Richard, and a few days later Richard was on the phone. I'd done a little homework on his literary efforts at the time, and I asked Richard if he knew that there was a file devoted to him (box 23, folder 11) in the Ginsberg Papers at Columbia University. He didn't, and I suggested that a field trip to New York City was in order to see what was there. Then we got down to business: I asked Richard whether he'd continued to write poems after the book was first published, and he allowed that, yes, he'd written some over the years. Then let me have I look, I said, and if they work let's add them to the original text so that we can make a book of forty-eight rather than the original twenty-eight pages. (It's all printer's math.) He growled a little at the thought of having to dig through his files to find them, growled a little more when I asked him to send the original artwork for some of his calligrammes, growled a little more than all that when I asked him to close a new edition with a little reminiscence of the life of his book from those days in San Francisco until the present. There was good reason for that, I thought: When I worked with Ed Abbey on a twentieth-anniversary edition of *Desert Solitaire*, I asked for just such a reminiscence, and the resulting essay, I think, is one of the best things Ed ever wrote.

Richard set to work, even writing a couple of new poems that were just as strong as I'd hoped, while I scanned the original book to get a working text started. In time the art arrived, then new poems ("Gaia Resets the Anthropocene," indeed), then photographs—for Richard, of course, was a fine photographer on top of everything else—and then, well, no more growls, but instead a quiet and courteous succession of questions and counterquestions about how to make the book known, always the most difficult part of any publishing

## LITTLE TURTLES

The mist clears.
fleshy fields
of yellow leaves
surround
steely desert ferns

the Sun rises
w/ you w/ me

The mist clears.
little turtles pour out of the sand
beckoned to the sea
by brightness
they disappear
in
swampy mangrove
root
refuge

fiddler crabs scuttle sideways
across manta-ray holes
in the lagoon
drained by the moon

Gorgonian corals.
sand dollars
touched by red tide
and sponges like soft brain coral
bleach brittle
at
the abandoned fish camp
flat turtle-barnacles
like empty eyes from the sea
staring out at the desert

a
Lion skull
& eggshell pottery sherds
scattered on the strand
w/ a green prickly-pear

"Little Turtles," from Richard's book of poetry, *Dark Horses and Little Turtles*, 1974. Richard Felger archive.

project. I published an electronic edition of the book, as the copyright page says, "in the COVID summer of 2020." And then, because publishers are supposed to do this, I asked Richard what he'd like to follow up with, suggesting that he turn to a memoir that would account for his many lives.

Alas, the one life that he had was swiftly coming to an end—and only days before the printed version of *Dark Horses and Little Turtles* arrived at his door. I'm grateful for having been given the opportunity to work with him on the book, as I'm grateful for having had the pleasure and privilege of knowing Richard Felger for so many years. If heaven is the land of our dreams, then the one Richard is now exploring looks just like the Sonoran Desert, dark horses on every ridgeline, little turtles in every arroyo.

26

# A MENTOR AND A TURTLE MAN TOO

Jeffrey Seminoff and Wallace J. Nichols (1967–2024)

AS SCIENTISTS, none of us have gotten to where we are in our careers without the advice and encouragement from key advisers along the way. Among those mentors are always one or two whose appearance on the path is transformative, pivotal, life saving, or all three. For us, one of those figures was Richard Felger. While many of his longtime colleagues and friends called him Dick, we always called him Richard—perhaps a generational thing, or perhaps our way of showing deep respect. After all, by the time we met him, Richard was already a renowned author and botanist, and we were just starting our graduate careers at the University of Arizona. And to know him was to always be encouraged to "go for it."

One of the most important examples of that was in 1992, when we were just starting our MS degrees under the tutelage of Don Thomson. At the time, we both knew that we wanted to study sea turtles in Mexico, but after a much-anticipated meeting in Thomson's basement office on the U of A campus with Professor Emeritus John Hendrickson, our plans were called into question. Hendrickson—then retired, but with huge sea turtle credentials and a pioneer in the field—had told us our graduate research plans were too risky. Granted, this was during the absolute trough in Mexico's green turtle (*Chelonia mydas*, a.k.a. black turtle) population and the height of the black-market turtle trade. "Not enough turtles!" he said. The green turtle population had been in steep decline for decades and was considered commercially and ecologically extinct by experts. Furthermore, working directly with current and former turtle hunters (a key part of our plan) was not encouraged, as the complexities of the illegal trade were considered academically challenging at best, physically dangerous at worst.

We both remember leaving that meeting somewhat deflated, but shortly thereafter we met with Richard—the three of us would meet often and talk Mexico, turtles, and life—and he insisted that we not buy into the negativity, encouraging us to follow our original plan. And thank heavens we did! As a direct result of this, we launched what we called Proyecto Tortuga (now known as Grupo Tortuguero de las Californias) and set off on an expedition to visit every single sea turtle nesting beach in Mexico. As we rolled out of Tucson in an overloaded 1975 Toyota Land Cruiser, Richard saw us off. He admonished us, as always, to take meticulously thorough field notes and to photograph everything.

Wallace J. Nichols (*left*) and Jeff Seminoff in Richard's driveway in Tucson, their last stop in town, prior to embarking on a 6,000 km circumnavigation of Mexico to study the country's sea turtles, 1992. Photo by Richard Felger, courtesy of Jeffrey Seminoff.

Today, the green turtle population in Pacific Mexico is in a major recovery phase thanks to years of conservation at beaches and in the water, with annual nesting numbers not seen since the 1960s. Our naïve and "risky" circumnavigation of Mexico helped to catalyze a community science–based movement that remains the backbone of sea turtle research and conservation in the region. And in a very real way, Richard had a key role in the process.

What many didn't know, or didn't remember, about Richard was that nearly two decades earlier, in 1976, he along with Kim Cliffton and Philip Regal published one of the most important sea turtle papers of its time: an article in the journal *Science* titled "Winter Dormancy in Sea Turtles: Independent Discovery and Exploitation in the Gulf of California by Two Local Cultures." To say that we had memorized every detail of that paper, as well as his other sea turtle–themed contributions to the literature, wouldn't be an exaggeration. Not only was the *Science* paper the first to describe the unique overwintering strategy of green turtles in the Gulf of California, but it also sounded the alarm about Gulf hooka divers' indiscriminate and unregulated take of a most-vulnerable species. This was also around the time that Richard had attended a meeting in Bahía de Kino, Sonora, when he came across two green turtles in their beachfront hotel's kitchen awaiting butchering. As he recounted in a 2019 email, "I got a few guys and we went in, picked up the turtles and everyone followed us down the big sand dune and we launched them into the sea." Richard was a turtle man, just like us!

We have often pondered why Richard took such a liking to us. He always seemed so busy with his book projects, plant collecting, and expeditions, but never too occupied to spend time with the two new U of A graduate students that would make weekly pilgrimages to his cactus-gardened house in Tucson's Sam Hughes neighborhood. Maybe it was our youthful naïveté and exuberance, or perhaps because as budding desert-bound sea turtle biologists starting our research in the Gulf of California, we were a novelty of sorts, giving Richard a chance to talk turtles and the Gulf, to dust off old tales that he wanted to tell and we needed to hear. Maybe he also recognized that without his guidance, we'd likely wreck ourselves.

Decades earlier, Richard had spent enormous amounts of time learning and living with the Comcaac (Seris) in Sonora—whose culture had relied on turtles for spiritual foundation and sustenance since prehistory—studying their traditional uses of plants and animals en route to publishing *People of the Desert and Sea* with Mary Beck Moser. It is in that book, for example, that the world learned about the encyclopedic knowledge of the Comcaac regarding all things sea turtles, whether it be the creation stories about the *moosnipol* (leatherback turtle, *Dermochelys coriacea*) or the incredible diversity they recognized in all the local variants (ten ethnotaxa!) of *moosni* (green turtles) found in the waters of the Canal del Infiernillo. It's no exaggeration to state that much ethnobiological knowledge of these critically endangered species would be lost if not for these collaborative works. And the current generation of young Comcaac sea turtle conservation biologists can trace its academic lineage directly through Richard's curious mind.

In this sense, we can understand why Richard loved to talk turtles, and we were fortunate enough to come along at the perfect time to share in our mutual interest in turtles and Mexico.

Like many of the Gulf of California "old timers" we knew, Richard's house was half library, half museum, surrounded by rare and exotic succulents and filled with artifacts, classic tomes, and vintage field gear. In addition to our conversations in this mini-museum, we would have opportunities to meet just as rare and exotic explorers, travelers, and scientists who would be staying at his house: Kim Cliffton, Lloyd Findley, and Tadeo Pfister, to name a few. In hindsight, our regular sessions at Richard's resembled meetings of Tucson's unofficial sea turtle fan club, with Dr. Felger our president.

One highlight of our young careers as biologists was when Richard invited us to co-author a chapter with him for Oxford University Press's book *Biodiversity, Ecosystems, and Conservation in Northern Mexico*. Our chapter, "Sea Turtles in Northwest Mexico: Conservation, Ethnobiology and Desperation," was a wide-ranging summary of the history and status of sea turtles along the coasts of Sonora and the Baja California peninsula. He taught us to be better writers, aggressive deleters of unnecessary words, but to also include the insightful observations that could be the seeds of inspiration to readers in the distant future, just as his overwintering paper had served for us. We were quite proud of that contribution and enthralled by that early taste of the writer-scientist life.

Another Felger-related early turtle career milestone was in 2002, when we were asked to co-host a one-day mini-symposium on the famed black sea turtle, on the first day of the week-long International Sea Turtle Symposium (ISTS) in Miami, Florida. At that time, mystique and curiosity about the critically endangered black turtle of Pacific Mexico were at a peak. Almost no one in the "Western" scientific turtle community had ever seen it before, and much of what they knew about it had likely come from Richard's writings. Although decades removed from any official turtle work, he still was among the world's experts in this peculiar species. We both thought that any Black Turtle Symposium would not be complete without a presentation by Richard. And so without hesitation we invited him to make the cross-country trek from Tucson to talk turtles with us one more time, along with a thousand of our turtle friends. He resisted just as we anticipated, claiming that he'd been out of the sea turtle business for decades. Yet we convinced him to come anyhow and to share his raw experiences and adventures with an eager new generation of sea turtle people. That he did: taking us on a science-filled storytelling ride populated with scanned-slide images and colorful recollections. That talk still stands as one of the best in ISTS history, in our opinions. But what happened next is burned into our neurons.

Dr. Rene Marquez—the recently retired head of the Mexican government's sea turtle research in Mexico, and for decades the most powerful (and feared) man in Mexico's sea turtle community—took the microphone to pose a question to Richard. "What do you think of the conservation award given to Antonio Suarez?" Rene asked. Suarez was a businessman widely credited with the decimation of both sea turtle and tuna fisheries under the guise of sustainable use, whose infamy exploded with Tim Cahill's sea turtle slaughterhouse article, "The Shame of Escobilla," in *Outside Magazine*. Richard took in the question and stood quietly at the podium for what felt like a full minute. Then he leaned into the microphone and said three words:, "That is disgusting." And walked off.

As time progressed and Richard and his sweetheart Silke Schneider moved to Catalina, Arizona, then Silver City, New Mexico, for more quietness and space for horses, we would spend less time with him but maintained email and phone correspondence. His messages were feisty, fine gems, ever guiding us along our career paths. One of the last emails from him, with his life force flickering, lamented that *all* he had time left for was "to finish the details of the few thousand pages of books with publishers" *and* finish writing "two other fun non-science books." His passion and productivity were legendary, and lest we get to feeling too special or unique, we know that he was a world-class mentor to legions and a mensch to all.

27

# THE MEETING

Gil Gillenwater

"HAVEN'T YOU HEARD, Gil? Jojoba is the next big thing!"

I grew to greatly regret hearing these words. The only positive outcome to the years and thousands, hundreds of thousands, of dollars that I and my investors would sink into the commercial cultivation of the jojoba plant was that it introduced me to a lifelong friend—Dr. Richard Felger.

It was June 17, 1990. I had just landed in Asunción, Paraguay, and it was hot. I was there to attend the Eighth International Conference on Jojoba and Its Uses. At the time, the plant appeared to have a bright and profitable future. And besides, the IRS was offering a three-to-one write-off for every dollar invested. So it was a "no lose" deal that my investors and I somehow eventually lost money on. But that's another story.

The story I want to tell now has me sitting in an undercooled auditorium in Asunción. It was like I had low-grade malaria as I struggled to listen, through underpaid translators, to foreign botanists and plant scientists drone on and on. They expounded in polysyllabic mouthfuls about triglyceride fats and oils being easily hydrolyzed and oxidized as well as free carboxylic acids, hydroperoxides, phosphatides, chlorophylls, color bodies, and malodorous low-molecular-weight carbonyl compounds and other equally fascinating, but to me nonsensical, discussions.

Couple that with the fog of jet lag, and I was proud to make it to 10:00 a.m. before I bolted. Bursting out of the auditorium, the dazzling sun blinded and disoriented me. Blinking helped, and soon I could make out large shapes. Then I saw it, the Lito Bar. It was a sign from Buddha. As I entered, I was blinded again, this time by the lack of light. Groping for a bar stool, I ordered a beer. I left the choice to the bartender, and he soon plopped a 24-ounce ice-cold Paraguayan Pilsen in front of me. Trance-like, I marveled at the beads of condensation as they slid down the tall bottle and pooled on the bar. Paraguay is hot in June.

The beer tasted surprisingly good, being that I'd never heard of it before. On to my second, I glanced down the bar. Wouldn't you know it, there was only one other patron, and he was a gringo as well. He nodded to me and a light discussion ensued. His name was Richard, and he was, likewise, in Asunción for the jojoba convention. He had become equally disenchanted with day one and, quite by chance, had sought refuge in this nearby bar. He told me he was from Tucson. "Tucson," I said, a little too loud. "Hell, I live in Phoenix and I drive through Tucson all the time." He didn't seem that impressed. "Really," he said in a dead-pan voice. "Why do you drive through Tucson?" Moving down next to him, I replied, "I travel to Kino Bay a lot and it takes me right through Tucson." Not waiting for a response, I asked him if he'd ever heard of Kino Bay. "Yes, I've heard of it," he said flatly as he took a long draw on his beer.

Ordering more brews for the two of us, I dove headlong into a one-sided conversation about how there was a tribe of Indians near Kino called the Seris. "Have you ever heard of them?" I asked. "Yes, I've heard of them, " but before he could finish his sentence, I went into rambling detail on how I went fishing with them out to Isla Tiburón in their pangas, and that I traded with them for ironwood carvings and handmade baskets.

By this time the third 24-ounce Paraguayan Pilsen was kicking in, and I was on a roll. I told him how lucky I was to be one of the few non-Indians to actually hang out with the Seris and on and on. He feigned interest.

Soon enough it was time to get back to the hotel, so we said our goodbyes. Before we left, he asked me for my mailing address. This was pre–cell phone, and I thought the request was a little odd. We saw each other a couple more times, but then I rented a car and drove off to see Iguazú Falls. I never thought I'd hear from this guy Richard again.

I'd been home a good month when this package arrived at my front door. It was heavy. I lugged it inside and set it on a table. Unwrapped the shipping paper, and a voluminous book appeared. "What in the world is this?" I wondered as I looked down at the title: *People of the Desert and Sea: Ethnobotany of the Seri Indians*, co-authored by Dr. Richard Stephen Felger.

"That son of a bitch!" I said to no one in particular. He sat there straight faced and listened to me babble on and on about the "almost a week" I spent with the Seris, all the while knowing that he had lived for months with them and written the ultimate 454-page treatise on the tribe.

From that moment forward we were fast friends. Over the years we kayaked the Gulf of California, climbed as "Bad Ass Botanists" in the Sierra Pinacate, hiked in the Goldwater bombing range, explored the *tinajas* of Nacapule and Barajitas Canyons in Sonora, rafted the Gila River at flood stage, pressed plants in the Seri village of Punta Chueca, closed down the Little Toad Creek Brewery in Silver City more than once, and shared other adventures too many to remember.

Richard was one of those unique individuals who knew exactly what he wanted to be at eight years of age—a botanist of the Sonoran Desert. There will never be another like him. His sardonic sense of humor had you hating him and loving him all at the same time. Even today when I'm hiking in the Sonoran Desert, I can still hear my botanist friend Richard shouting enthusiastically, "New find! New find!"

28

# RICHARD, MY FRIEND, MY HUSBAND, MY SOULMATE

Silke Schneider

RICHARD FELGER AND I met in 1994, shortly before his sixtieth birthday. He was giving a poetry reading at the Biosphere 2 Earth Sciences research facility in Oracle, Arizona, where I was working. Though I wasn't into poetry at that time, I loved Richard's. His reading captivated me, and I loved his presence, filling the theater space at Biosphere 2. I felt the presence of a remarkable human being up there on the stage. I had no idea who Richard Felger was, until some of my friends in the audience mentioned his work with the Seri Indians and how impressive he was to be such a poet and scientist. Somehow, we connected that evening, and that connection became strong and stronger over our twenty-six years together, until the night Richard passed away in my arms at our home in New Mexico.

After Richard's reading, our mutual friend Linda Leigh brought Richard over to my little trailer home in the Tortolita Mountains. Linda had worked for Richard in the past, and now she was a member of the Biosphere team, where we worked together and became good friends. I was employed at the Biosphere (actually I was the first candidate for the Biosphere 2 closure), but I also had a project at my home, breeding and raising emus. Linda brought Richard out that afternoon to witness a couple of emu chicks hatch. I still have that pair of emus today, Murl and Fred.

Richard happened to leave his little "botanist hat" behind that evening, so in a few days I delivered it to his house in Tucson. Richard opened the door a tiny bit, and I passed his little hat through the opening. Richard thanked me but said he was busy writing and couldn't be disturbed. We laughed for years about that scene, which Richard denied with a "who, me?" smirk on his face.

But we started phoning each other, and I remember asking him, "Would you sometime like to go out to breakfast?" He said no. I told him, "That's all right, Richard." But he explained why he couldn't by saying, "That's my best time to work. In the morning." "Well, how about dinner?" "Oh, that'd be great." We started getting together for visits over dinner in town, exchanging stories about our lives. We'd talk about our families and how we grew up and our plans for the future.

Richard's family is Jewish, and my family is German. One story that Richard shared with me was about one of his relatives, who was a well-respected pharmacist in Germany. When it got too dangerous for Jewish people in Nazi Germany, his relatives and family moved to the United States. Not liking his pharmacy work in the United States, they returned to Germany. Sadly, they were never heard from again.

Richard and I talked about our many adventures, including my roundabout story of how I came to be at Biosphere 2. In 1977, when I was eighteen years old and just out of high school, instead of going to college as my parents wanted, I ran away and joined a traveling circus. A year later I left the circus in southern France, lived on a houseboat in Cassis, and studied French literature in Aix-en-Provence. That's where I met up with an American experimental theater company on tour from its home base near Santa Fe, New Mexico, and whose director, John Allen, invited me to join them. In Santa Fe, after the tour, I fulfilled my dream of riding a horse along Old West trails. After nine months of training, I rode my Arabian horse 1,200 miles on the old Santa Fe Trail to Dodge City and then onward along the old Chisholm Trail to Fort Worth, Texas. Thanks to that ride I became a full member of the Long Riders guild and a Fellow of the Royal Geographic Society.

After that ride, I accepted a job as a jilleroo (cowgirl) on a cattle station in the Outback of Western Australia, where I lived at a stock camp in the bush three months of the year, managing the thirty or so horses needed for muster (round-up), and working closely with Aboriginal stockmen. It was the adventure of a lifetime. The spread was owned by Ed Bass, an American and a friend of John Allen, and both were deeply involved in the planning for the

Biosphere 2 project. In 1984, though still in its design stages, I was offered a position at Biosphere 2, and I eagerly accepted. Ten years later, after participating in the completion of the project and working there as an animal systems manager, I happened to meet Richard Felger at his Biosphere 2 poetry reading, and my life changed in big ways.

During our early "get to know each other" dinners in Tucson, Richard opened up about his upbringing in California, his passion for botany, life, and science, and his concern for the future of the planet. Richard was open minded, not narrow or linear in his thinking. My circus, travel, and Biosphere 2 adventures, my horses, love for animals, emus, cats, dogs, fit just as well into Richard's world as his very precise, scientific, fact-driven work. Richard was so expansive in his thinking, his interests, deep knowledge, and broad personal experience.

I've always been a person who doesn't mind being alone, and I've traveled mostly alone. But with Richard this was different. We kept our own space yet shared our lives, stories, and passions. And Richard had traveled a lot. He had been to Israel and Moscow, to Africa, to India and Mexico. He knew the cultures, the people, the places.

Our mutual friend, Linda, who introduced us could not believe how deeply Richard's and my relationship was developing. She'd say, "You are *so* different! I can't believe you're still together." Yes, so different with quite different backgrounds, yet we always shared very deep common values, and deep, deep appreciation for life—our own lives, life on the planet, and a big gratitude for having the life we enjoyed, and for having found each other. Richard embodied these values without many words. I would feel our common values of "oneness" with life and nature, the animals and plants, mountains and oceans, people and cultures. We could be on a field trip, or camped in the Australian Outback, or walking the dogs, or just in the garden and be in awe of the beauty around us, experiencing it all without saying a word.

Before I met Richard, even way back in Germany as a teenager, I had and still have friends who like to talk a lot about philosophy and life and death and other spiritual subjects. Richard wasn't one to talk about anything like that, but he expressed it through his poetry. He didn't express a creed, and he didn't call himself spiritual, religious, or even agnostic. He talked about nature, nature, nature. He felt a deep connection to the universe, and we shared being in awe of the universe and life. Where did it all come from? And all these forces that empower nature. I don't have one god. Richard was like that too. He didn't worship a god or embrace religion, and he wasn't a Buddhist, but he lived in awe of it all. Life, nature, and evolution: Where does it come from, and where is it all going?

He had an incredible passion for writing, especially for poetry, as if words were seeds to grow. The stories he told! And his poet friends were as fascinated by him and his knowledge of the desert as he was fascinated by them as writers of poetry. He said that he used to be heavily involved in population biology, and he was on TV and interviewed. But becoming a famous person on TV was not his ambition, which is where it might have gone if he had continued. Still, population and sustainability focused his vision.

He had a deep love for children. When we stopped someplace in Mexico, he talked to every one of them. That interested me because I work with little kids. When I got my first job in Oracle, Arizona, as a kindergarten teacher, the principal interviewing me for the position said, "I see you studied animal science. Why do you want to be a kindergarten teacher? Wouldn't you like to teach science in a higher grade?" He was worried that I might quit after starting the year. I told him, "No. Kindergartners are the best scientists," and they really are. When I see those little kids, how they see, and how they touch, and ask amazing questions, I think Richard connected with them on that level. He had a child's curiosity, and he never lost that enthusiasm. He wasn't trying to sell them anything, and he wasn't saying, "Don't touch." That respect for children was one of our deepest connections.

One time a couple of families from my school wanted to bring their kids out to see my little pony, Frankie the Horse. So they came, and Richard joined us. We went into the greenhouse with the little kids, five- and six-year-olds. The adults were admiring the flowers, and do you know where the kids were? On the floor looking at and touching the gravel. And they were fascinated. Richard fed their curiosity and encouraged them—that's how he got young people interested and involved. He frequently did field trips for the native plant society and did lots of teaching on the way. He even came to my school to talk about plants and roots. The kids loved it.

On the teaching part, one of the things that I'll always remember is how much time he dedicated to teaching me, especially reading drafts of my papers. At that time I was giving presentations at Rare Breeds meetings in the United States as well as internationally with Rare Breeds International. He would read them meticulously and give me the most detailed notes draft after draft. He showed me the ropes and made me realize the time that writing takes. The devotion, that commitment, is priceless. Too, he helped graduate students with their theses or dissertations. He really loved doing that, and he felt appreciated.

Richard's father, Louie, was a physician, and the family lived at Malibu in the summer where Richard explored the beach as a child. That's why he said, "I was a marine biologist before I became a botanist." His dad was a very dedicated physician, and Richard said often the family would sit down for dinner and his father would be called away to a patient's home.

Sadie—his mom—and Louie always had employees. They had a nanny, and the nanny lived with them. She even went on their vacations to take care of Richard and his older brother, Danny, whose name was actually Dan, but we called him Danny. Richard loved to tell the story about the family going on a ship to Europe for a vacation, and the nanny went along. Richard joked how the boys pretended to be seasick, which induced the other passengers to throw up and not want to eat. That way the boys could get all the food for themselves, especially the caviar.

Richard and his mother, Sadie, September 2008. Courtesy of Silke Schneider.

When I met Richard, his father had already passed away, so I never knew him. Richard told me that his father was a famous surgeon and doctor, but he had to have some procedure done on himself, and the anesthesia partially paralyzed him, so he ended up in a wheelchair. Shortly after that he died.

But Richard was very attentive to and protective of his mother and called her frequently. Every time before he crossed the international border, his first phone call was to me and his second call was to her, saying that he was headed out on a trip or that he had arrived home okay. A long time ago my mother told me that when you meet your future husband, watch how he treats his mother. That's how he'll treat you.

After Louie died, Sadie remarried a famous lawyer named Martin Gang. When I met Sadie, she had photo albums of Richard's dad and stepdad on her coffee table, and she'd always say, "I was a two-time winner."

Martin Gang was a lawyer to Hollywood stars and a supporter of Jewish causes, including Cedars-Sinai Hospital in Los Angeles. About the time I met Richard, Gang got dementia or Alzheimer's. Sadie couldn't take care of him, so Gang went to an assisted living home, but she would go every day and visit him. Richard was concerned because she was still driving. Eventually he convinced Sadie to have someone else drive her.

When we went to see her, she was already in her eighties and nineties. In his will, Gang took care of Sadie for the rest of her life. She had a nice apartment in Beverly Hills, though Richard called it a "tenement," and she had people helping with everything at her apartment—somebody to cook or to clean or to tend her plants on the balcony.

I knew very little about Sadie when we first went to see her. Richard said, "I really would like for you to come meet my mother, but don't be shocked, because she lives in Beverly Hills. My brother is in real estate, with lots of apartment buildings, some even in Tucson."

When we first met in person, Sadie knew very little about me, but she had already bombarded Richard with questions like "Does she know how old you are?" and "Have you told her how many times you've been married?" She did know that I liked horses, and she told me that as a child in Montana, "We would go for rides in a horse-drawn buggy, and one time the horses ran off and the nanny protected me, or else I would be dead." She was only three years old when that happened. And Sadie would tell me lots of stories about how in the early days, the family couldn't be members of the country club because they were Jewish. Sadie had a big heart and an amazing memory.

We used to visit Sadie at least two or three times a year, including Christmas. I said to Richard, "I thought your family is Jewish. How come Christmas?" He said, "Sadie likes to celebrate everything." We often went for Thanksgiving and celebrated her birthday at the same time, even though it was in September. And the whole family would be there, four generations, later five generations. Richard and Danny did not have much in common, except both really cared for Sadie's safety and well-being. Right after she passed away, at the age of 104, we lost touch with the family. Then Danny died from an aggressive type of skin cancer, which is why Richard was always so worried about protecting his skin.

The Felger extended family gathering in celebration of Sadie's 103rd birthday, September 2008. *Back row, left to right*: Silke Schneider, Richard Felger, Carolyn Felger, Dan Felger, Brian Dwyer, Danielle Dwyer, Hailey Felger, Brad Felger, Donivon Hailey. *Middle row, left to right*: Tracy Rosberg, Sadie (Felger) Gang. *Front row, left to right*: Valerie Felger, Holden Rosberg, Mickaela "Mikey" Madison Rosberg (holding Ivorie Felger), Miles Rosberg. Courtesy of Silke Schneider.

Richard was very successful at what he did and lived an exemplary life, but Sadie didn't always appreciate him or what he accomplished, or even understand what he did for a living. She couldn't understand why he didn't have any money, so she might give him $5,000 toward buying a new Isuzu. When Richard had a new book, he would take a copy for Sadie to see, but there's no way she could ever understand the effort and achievement it represented. She'd ask him, "Are you going to get any money for it?" For her, that was the first thing: When are you going to make any money?

One time Richard won a special award, but he didn't even want to say anything to her, because he didn't feel appreciated. Sadie would say, "Richard has done this and he's got a book out." And that was it. But I think she was proud of him, so I told Richard, "She doesn't understand what you do, but she is proud of you. You know that." She was very happy when we were together. She was warm, and I always felt welcome. And she asked me about my horses. She even knew their names. I said, "Richard, you need to tell her more of what you do." And he said, "Oh, she won't understand." I told him, "That goes two ways."

It was the same in Germany with my parents. I thought there's no point in even talking with them about my jobs and my travels. After high school I was supposed to go to a university and find a good husband, one making good money and with excellent credentials, preferably a high position in the army like my father, and we'd have kids, and all of that. But that wasn't my path, and my parents could not understand it at all. But they did later. When I joined the circus, it was the same thing. They didn't tell anybody. They just said, "Oh, she is abroad doing studies," so nobody would ask. They were super embarrassed, and I felt bad about it. But on the other hand, I needed to follow my dreams. I told them, "I'm always going to be here for you, but I'm not going to live your life." After the circus and after I had lived in all these different places, my parents and I became good friends. And when they came to America and met Richard, they finally did see my very happy life.

Richard lived from grant to grant. As much as that must have worried him, he still focused on what he was doing. When our income was tight, he shrugged and just said, "We'll get through this. We'll keep moving." I greatly admired him for that. Although my jobs teaching and selling real estate helped, they were not enough. Richard had long relied on H. B. Wallace for grant money, but when he died, his family took over for Mr. Wallace, and funds were limited. Richard just said, "Don't worry about it," which is a great attitude, and somehow we always made ends meet.

Richard might go to somebody's yard or go to a friend's party and the first thing he said might be, "That plant needs water. That one needs to be repotted." At our home it was the same thing. I can see how people might take offense at that. Or if you acted interested in his project, he might ask you to help. During a party at our house, he might take advantage of the situation and say to someone, "Okay, you can do this and that and that and that." I've seen him do that with people, but I read it a bit differently. He was always working on getting his projects done. He was thinking, "We have some labor, let's get these people busy, they'll learn something, they'll get it done." He would have never held it against them if they said no, but I can see how maybe some people misread that. He meant it as a compliment, an invitation.

When we lived in Catalina, we would hold little gatherings for dinners or host out-of-town guests for a few days. Richard would sometimes disappear, and then everybody would wonder, where's Richard? He was just upstairs reading, and he'd say, "Don't worry about me. I'm just reading." Or he'd say, "I've got some work to finish." I had to explain to him that a lot of people are not comfortable when you leave during dinner. He didn't mean anything by it; he just had other things to do, or he wanted to go to bed.

He had a very positive and pleasant disposition. He was always even tempered. He never got mad, and he seemed to tell me everything that was on his mind. He had his work, his purpose, his goals. He was always very generous and very kind. And Richard was modest. He earned his PhD because he was interested in the subject, not to be a "Doctor Felger."

He was never sentimental, which also attracted me. He was matter-of-fact. He would say things that I thought were very poetic, but he didn't write poetic notes to me. In fact, when I worked as a vet tech at a veterinary clinic in Tucson at the beginning of our relationship, all the girls in the office got roses for Valentine's Day. From Richard, I got a cactus. And I loved it.

Shortly after we met, Richard was diagnosed with prostate cancer. I was freaking out. I was only thirty-five or thirty-six years old, so I didn't know many people who had cancer. Richard was amazing. He said, "Let's just deal with it. Let's just check out the options and go forward." He always did well for himself. It was the same when he started having trouble with his health three months before he died. He said, "Okay, let's just deal with it." And he did, until those final two weeks. But he was so happy to be here at home, and basically we were hopeful until the very last week. The main regret that he expressed was we didn't have more time to spend together.

He was always very health conscious. And he exercised and went for hikes. When I met Richard, he complained about his back, because he was sitting a lot while writing. But as he walked more, he improved. Even friends were asking how could he stay in such good shape when he's always sitting at the computer. He would mix work with play, and he was walking our trails and taking the dogs up and down our steep long driveway.

Richard appreciated nutritious and tasty food. One of the stories he'd tell frequently was when someone asked him to dinner at their house, they'd teasingly say, "Come to dinner. It's just like mother's cooking." He'd joke back and say, "No, thank you. My mother's cooking was horrible." Yet when we ate at Sadie's house, the food was sumptuous and divinely delicious. But *she* didn't cook it.

At home he would graze in between meals, and when I was teaching, I ate lunch at school. Then we'd have dinner and coffee together, but even scheduling dinner sometimes was hard because I would like to work outside until dark. I like to eat a bigger meal after I'm done, but he liked to eat early. So we compromised. We'd eat together a little bit and visit, so he could eat early. He would have exciting news, something that happened, a brand-new grant was in the works, or a book or research paper was accepted for print. Then I finished my chores outside, and he read or worked in his office.

I think if somebody has a good relationship with a former girlfriend, that speaks well for them both. And I've met a few of Richard's former girlfriends, and I talked with one of them who once lived in Patagonia, Arizona. By then she had another boyfriend, and they had married, but she spoke highly of Richard. She said, "I'm so glad you're together and he found you." That says a lot. It's a sunny day when people still get along. That was true even in Richard's family. For Christmas at Sadie's home, Danny's former wife was there at the party along with his current wife, Carolyn. Danny was very respectful of his former wife and of Carolyn. I thought that was so cool for them, the kids, the grandkids, and Sadie.

Apparently, Richard was married two times before me, but I never asked much about it. He was very young, first time and second time, but of course I was curious if he had kids. I asked, "So how come you always got divorced so fast?" And he said, "Oh, she only wanted to watch TV." Or, "I worked all the time and the wife didn't like it."

I can completely get it. Richard and I were made for each other, but it's not your ordinary relationship. We didn't go out together on Sunday to have breakfast. He did his thing, hiking and botanizing with his friends, and that worked fine for me, because I liked to do my thing, riding horses with my girlfriends. And then of course, we did our field trips, dinner parties, and many social events together.

When we started off, we shared all the same spaces in the house. But he was reading all night and working as I was sleeping or trying to sleep. And then there was his computer and the phone ringing. I thought, "You've got to be kidding; this is not very romantic." So I tried to clean up the common space, which was driving both of us nuts, and then I tried to cover all of his scattered books and papers with pretty Mexican blankets. Seriously, he didn't want his books and papers moved. I was putting blankets over them so it all looked less cluttered and nice. Well, that's crazy. Why do that to each other? I learned that Richard thought there was something wrong if we didn't share every space together, but his space seemed everywhere, and I've always liked my own space. Together we decided on a solution. He chose one end of the house, and I chose my end of the house, and then we got together for dates. That was a beautiful thing, and it worked well for us.

After Richard's sixtieth birthday party—he told me he was thirty-nine—I returned to Australia during my summer break from college to work at Quanbun Downs station, where I had worked in the early 1980s. Richard arranged to visit me there after the round-up was over. We met in the town of Broome and traveled the Outback along the Gibb River Road in a little SUV I had rented for us. Richard was elated to come across an amazing cycad forest, lizards, snakes, flying foxes, amazing birds, crocodiles, and overwhelming natural beauty without a human soul in sight. And we discovered that we were a perfect match for each other.

Shortly after our Outback adventure, we moved to a country home in the foothills of the Catalina Mountains north of Tucson, and we were awed to share our lives with all the plants and animals that we found there and that we brought with us. We were both amazed, enchanted, and even amused at the different yet so similar worlds that we now shared. I had no idea it could take hours to explore for plants along a short trail in the mountains and deserts,

Richard and Silke with their rented SUV in northwestern Australia, July 1995. Richard Felger archive.

Richard and crocodile, northwestern Australia, July 1995. Richard had a special fondness for crocodilians. As a teenager, he kept alligators as pets and once tried (unsuccessfully) to keep a crocodile. Richard Felger archive.

nor was I aware of all the plants and treasures that could be found. On the other hand, Richard could not believe that we horse riders would trot so fast right past those amazing treasures. Attending plant meetings and attending horse meetings were equally amusing and broadened our fascination with our worlds and each other, so different yet so much the same.

We were thoroughly happy in Catalina with the house and the property. Richard did hike there because we could walk in the state park, but the heat got worse for him over the years. And his commute to the herbarium was getting longer by the day because of traffic. So we started looking at Silver City, and Richard really liked it because it had a university, which he needed, and an herbarium and the hospital medical facility. Plus, it has a very interesting diverse population and a fabulous climate, and it's not far from Tucson, about three hours to downtown or the university. He wasn't doing as much fieldwork or traveling then. Instead, he was shifting to writing manuscripts and finishing his projects. And he didn't have the administrative chores of Drylands Institute anymore.

So in 2007, Richard and I moved to Silver City, New Mexico. We found a beautiful home and ranchette bordering the three-million-acre national forest, with endless trails and beauty. We totally fell in love with our quiet home on the mountain just minutes from town. We shared our mountain with coyotes, bobcats, mountain lions, bears, deer, birds, lizards, and snakes, as well as my domesticated friends: the emus, chickens, turkeys, horses, dogs, and indoor cats that I brought with us. We lived right next to the Continental Divide Trail, and we could hike and ride horses straight out the door.

At home, Richard was able to do his work without a commute, with a fresh mountain breeze blowing through the window, and his elegant greenhouse a few steps from his office. We enjoyed being home, especially during the COVID years while I was teaching online from our home. For me "teaching online" meant live kindergarten lessons as well as recorded. Our school district had supplied all our school's students with iPads to take home during COVID, so I set up a corner "studio" in our home's hallway to teach my classes. Here I danced and sang, did puppet shows, and told or read stories. I even got Richard out of his office and amused him as he witnessed my attempts to engage five-year-olds via screen time. I also did "nature hunt" lessons in the forest around our house and lessons on how to get Frankie the Pony ready to pull his cart. Richard also would record some of my outdoor and ranch activities so we could send them to my class. Richard and I spent all our time together and truly felt happy and thankful for our life together.

From the age of eighteen to thirty-five, I had traveled and worked in many countries around the globe. In Silver City my life was to be here and to be a teacher, and I didn't have the constant urge to be on the road. It felt like I had arrived home with my soulmate, friend, and husband, Richard.

For a long time before Richard passed away, maybe even ten years before, he was already working on his final books. Richard worked from 4:00 in the morning until dark. He was always working. Even when he hiked and played, he was always saying, "I have to finish these books." He was healthy; I had never seen him sick with the flu or a cold. He took very good care of himself. But he would say things like, "I'm getting old, and I may only have ten or fifteen good years left." I would sometimes say, "Just take it easy. Let's just go out or on a vacation," but he didn't know the word *vacation*. He might go someplace for a reason, like to work or to see plants, but not to play or relax. Richard and I went only once together to see my parents in Hamburg, Germany, but the bigger reason he went was to see Hamburg's botanical gardens, where many of the staff knew of his work and recognized his name.

When Richard started writing about his adventures in the field, he gave it the whimsical title *10,000 Years of Field Notes*. Richard's calling was his scientific work, and at first he didn't take the project seriously; it was a side job at most. But I was utterly fascinated by these amazing stories, and I encouraged him to complete them. I loved reading his chapters as they evolved into this book. The fascinating life of a captivating, passionate, dedicated human unfolds when we dive into them.

I feel privileged to have shared our lives together for over a quarter of a century.

29

# A BOTANIST MAKES HIS MARK IN SILVER CITY

William (Bill) Norris

WHEN I ARRIVED in Silver City in 2001, I had never heard of Richard Felger, nor did I once hear his name uttered for the next four to five years as I went about my business as the new botanist at Western New Mexico University (WNMU). Then, in the mid-2000s, I received an email out of the blue from Richard, announcing that he was preparing to move to Silver City and thought it appropriate to introduce himself to the curator of the local herbarium (i.e., me). The email message, which was brief and deceivingly modest, pointed me to an attached file containing a list of his publications. I opened the document and was immediately stunned by its length: far more than one hundred literature citations—books, published papers, and so forth. Richard was clearly someone I would need to reckon with when he arrived in town.

Richard was a true eccentric who grew up in Southern California and lived for many years in Tucson, and it took him several years to find his place in the well-established botanical community of Silver City. I think that it was a surprise for him to discover that his well-earned reputation in Arizona as a botanical scholar did not automatically precede him when he moved to New Mexico—that not every botanist in Silver City knew of him or of his prestigious botanical career. That said, Richard quickly went out of his way to meet other plant-minded (and some non-plant-minded) people in his newly adopted home by joining and becoming active in several organizations, including the Gila Chapter of the New Mexico Native Plant Society, the planning committee for the biennial Natural History of the Gila Symposium series, a local hiking club, and a local writing group, and he soon became a familiar member of the Silver City community. He also worked closely with proprietors of several local native plant nurseries to give advice on selection and propagation of plant species that he felt have potential as future food crops on our warming planet. One could always count on seeing Richard wearing his broad-brimmed hat on Saturday mornings at the Silver City Farmer's Market, or occasionally working out at a local gym. I frequently heard Richard say that he loved living in our small but vibrant town and had no desire to ever travel to hustling, bustling Tucson again.

Richard never admitted to being retired, bristling at the notion and always considering himself to be an active botanist looking forward to the next, hopefully funded project. Soon after arriving in town he realized that he did not have command of the Gila National Forest flora to match his deep knowledge of desert plants. Richard thus took it upon himself to accompany several local botanists and plant enthusiasts (especially citizen-scientist Russ Kleinman and nature photographer Elroy Limmer) on numerous field trips into "the Gila" so that he could bring himself up to speed with this new flora. He loved almost all aspects of botanical inventory work—plant collection, pressing, identification, and so on. Like many botanists of his generation, he collected many duplicates of each plant with the intent that they be distributed to herbaria throughout the Southwest. Richard did *not* enjoy preparation of specimen labels, relying on his co-collectors to attend to this task or hiring others to prepare labels from his field notes. The Dale A. Zimmerman Herbarium at WNMU is now the repository of hundreds of plant specimens he collected in New Mexico.

It took a while for Richard and me to warm up to each other, but that we did after spending time together during a variety of botanical adventures for which the goal was to observe and learn about plants in the habitat du jour. Upon learning of my research on sedge taxonomy, he gleefully announced that he "never, ever, had to identify a sedge species in the desert." But he certainly knew desert plants. I recall a spring field trip more than a decade ago with Richard and several other plant enthusiasts to the dry foothills of the Tres Hermanas Mountains, just north of Columbus, New Mexico, on the Mexican border, just as a variety of desert herbs were emerging, and being impressed that he was able to announce without hesitation the scientific name of every single plant. It was on that day that I realized the depth of Richard's knowledge of desert plants.

Richard had an amazing eye for plants. I recall a day-long excursion with him to the Gila Cliff Dwellings National Monument (as part of a decade-long, recently completed plant inventory of the monument) in the early 2010s, and being amazed when he spotted a diminutive red-stem monkeyflower (*Erythranthe rubella*) growing on the thinnest of soil over rock adjacent to the main dwelling. I, with my twenty-plus-year younger eyes, had completely missed it. I also remember when Richard, in a town whose streets are lined with Siberian elms (*Ulmus pumila*), picked out a single and very similar Chinese elm (*Ulmus parvifolia*) tree planted on the WNMU campus that I had unknowingly walked past for years. While conducting fieldwork today, I sometimes wonder what other plants I might have overlooked that Richard would have spotted and brought to my attention.

Richard contributed to many botanical education and outreach programs during his Silver City years. He was frequently called on to give the keynote address at various annual meetings and conferences and made sure that his botanical acquaintances were aware of these invited feature presentations. He, Russ Kleinman, and I co-presented at least six plant identification workshops (e.g., grasses, composites, woody plants) for the local chapter of the New Mexico Native Plant Society, and we quickly discovered that we could all contribute substantially to these workshops without tripping over or trying to one-up each other. Participants in these workshops greatly enjoyed our team approach to presentation and how we prompted each other to offer up some botanical tidbit or another. During these workshops, Richard's common response to being handed a plant with which he was familiar was "Aaah, [plant name], my old friend!" His utter fascination with plants was always on display during any workshop or field trip that he led or co-led, and his enthusiasm was infectious.

It was not until after Richard's passing that I realized how much I learned from him that I pass on to my students while teaching botany courses at WNMU or leading field trips. Just about everything I know about ethnobotany I owe to casual conversations with Richard, including the culinary virtues of honey mesquite (*Prosopis glandulosa*)—"mesquite cookies, anyone?"—and also that Indigenous people would consume Emory oak (*Quercus emoryi*) acorns right off the tree. Richard also gave me a useful tip for distinguishing Emory oak from other spinose-margined oak species (e.g., scrub live oak, *Quercus turbinella*) by that little tuft of hair on Emory oak at the very base of the lower leaf blade at its junction with the petiole. During wildflower hikes that I now lead for the general public, I always refer to *Zuloagaea bulbosa* as Apache redgrass, a moniker that he coined for one of his favorite grass species that, in his view, have potential as a future food crop. As a floristic botanist, I frequently pass along Richard's knowledge that, in comparison to the relatively intact native flora of southwestern New Mexico, the floras of many other regions of the United States (e.g., southern coastal California) are often overrun by non-native plants.

When asked to introduce himself in some group situation, Richard would invariably say, "I'm just a botanist," with a twinkle in his eye and no further elaboration. I am not sure why he was so withholding in these self-introductions—false modesty? a desire to cultivate intrigue?—but I do know that he was more than a "just a botanist." Poet, educator, raconteur, outdoorsman, conservationist, and scholar all apply to him. Richard Felger definitely made his mark in Silver City.

30

# THE GREAT WILLOW ADVENTURES

Russ Kleinman

WHEN RICHARD ARRIVED in Silver City, New Mexico, he didn't know the local flora. He had a four-wheel-drive Isuzu Pathfinder, but he preferred not to drive off road himself. Since most of the places to botanize in southwest New Mexico are off the beaten and paved path, Richard needed to befriend someone who knew the local flora and could drive off road to the best places to look for plants. I believe I met Richard shortly after he arrived in town at a Gila Native Plant Society meeting at Harlan Hall on the Western New Mexico University campus. Harlan Hall was very convenient for Richard, since it was just a few miles from his new home. Harlan Hall also houses the Dale A. Zimmerman Herbarium, where Richard spent hours at the dissecting scope examining plants and making measurements to include in his writings.

Richard and I became hiking and botanizing companions very soon after we met. I had a four-wheel-drive, high-clearance Toyota 4Runner. I also knew where to find the most interesting trails and plants. Perhaps most importantly, I had detailed knowledge of the oft-times informal dirt roads throughout the Gila River basin. These are "roads" that Richard euphemistically called "textured." Our travel over them involved climbing over rocks at two miles per hour as often as not. These trips didn't bother Richard at all, as long as he wasn't the one doing the driving and there were interesting plants at the end of the road.

Kevin Keith and his wife, Sarah Johnson, lived in Cliff, New Mexico. Sarah was the editor of the statewide newsletter of the Native Plant Society of New Mexico. Kevin, who is an excellent botanist in his own right, had explored all the wilderness in southwest New Mexico, especially those places he could access better than anyone else closer to Cliff and the Arizona border. Kevin interested Richard and me in a day trip to the Apache Box, a canyon right on the Arizona border. Kevin tried to prepare us for the hike by telling us in no uncertain terms that it was difficult. We weren't flustered—after all, the entire hike was only about six miles round trip. Sure, there were no marked trails to follow. We were used to that. Yet, nothing really prepared us for what was to come. . . .

On April 24, 2009, I picked up Richard early in the morning at his house, and we drove to Cliff to meet up with Kevin and Sarah. The weather was clear, and the temperature was perfect. The four of us then headed about twenty miles northwest to Mule Creek, with Richard and me in the 4Runner and Kevin and Sarah in the lead in their truck. The pavement ended at Mule Creek, and from there we turned onto Brushy Mountain Road, a dirt road that we followed for several miles. We then turned onto a fork that heads south and downhill to a locked gate, where we parked and donned our daypacks. Each of us was carrying about fifteen pounds, including lunch, several liters of water, rain gear to protect us in case of a change in weather, and emergency supplies. In addition, Richard carried equipment to collect botanical specimens of interest.

The first part of our hike was rather gentle and lulled us into complacency as we skirted private ranch property. Kevin was our guide, and he made it very clear that we could get to the bottom of the box only by one route. If we carried our path too far west, we would "cliff out" in Geronimo Canyon and not be able to get to the canyon floor. "Cliffing out" is that disappointing arrival at a vertical drop that can't be circumnavigated without traveling miles around it.

Not far into our journey, it became obvious why the Apache Box is considered a difficult hike. The ground is completely covered in potato-sized, rather smooth but pitted oval rocks of volcanic basalt. Each of these rocks is loose and free to roll under your feet. We all fell down many times. It was as if we were walking over fist-sized marbles. Miraculously, none of us was seriously injured.

After what seemed like an eternity over untrustworthy footing, we crested a rise. The view that greeted us was worth every bit of the hike. We were looking down into a canyon with four-hundred-foot-tall walls that were vertical near the top and undercut near the bottom. The floor of the canyon was cluttered with school-bus-sized boulders. There were hanging gardens of plants decorating the canyon walls. What a view!

The best part of all was that Kevin had found the proper approach to the canyon floor. We descended rather quickly and ate lunch along the creek, sitting on rocks at the east end of the box. Since I did the driving, Richard always made us lunch that consisted of the most delicious sandwiches. The vegetables were sourced from his own garden or from the Farmers' Market in Silver City, and the bread bought at the Silver City Food Co-op.

After lunch we headed downstream along the creek. Just a minute or two after leaving the site of our lunch, Richard was dancing gleefully around a tree. He had become very interested in the trees of the region, and this was one of the most special. He had found one of the first known specimens of *Salix bonplandiana* (polished willow) in southwest New Mexico. The young twigs are red, and the leaves are green above and white underneath. The bark of mature trees is heavily ridged. He took many measurements, including using an inclinometer to measure the height of the tree. As always, he asked us to stand at the base of the tree for scale. Richard was in his element here.

We spent an hour or two hiking over and around the huge boulders in the canyon before we had to start back for the vehicles and the trip home. I struggled to keep my footing, and I fought fatigue even though I was much younger than Richard and in good shape. Richard seemed to have no problem at all. Even though we were tired, the hike back seemed shorter than the hike into the box. Maybe we knew now what to expect along the route, or it could have been that we were so elated at what we had just seen and done.

On the drive home we discussed where we might find a similar suitable habitat for *Salix bonplandiana*. We settled on a future hike into the western reaches of Sycamore Canyon. The creek at the bottom of Sycamore Canyon flows only seasonally and is a tributary of the Gila River. It took us six months to finally undertake it. I wanted to make sure that we wouldn't be crossing private property, so we checked the maps of the region before deciding to make the hike.

Early on the morning of November 22, 2009, we left Silver City heading north on US 180 toward the towns of Gila and Cliff. After crossing the Gila River, we turned south onto Iron Bridge Road and eventually met up with Black Mountain Road, which we followed until it became nearly impassible.

After parking, we climbed over a fence without any "Private Property" signs and descended gently downhill into Sycamore Canyon. The canyon is beautiful, and this part of it is mostly unvisited by hikers. Botanists had probably never before laid eyes on it. Almost on cue, and nearly unbelievably just after arriving at the canyon bottom, we were staring at another *Salix bonplandiana*! Photographs had to be taken and specimens collected. Richard was once again in his element. We hiked farther west and found that the first tree was not the only one of its species in the canyon. We ate lunch in shade at the side of the dry streambed and discussed our good luck this day.

Richard and I went on many hikes together after our Apache Box and Sycamore Canyon adventures. We enjoyed them all, but most of them involved reminiscing about our trips to see *Salix bonplandiana*. I don't know if Richard ever saw another one of these trees after the ones in Sycamore Canyon. I'd like to visit Apache Box again someday, when I gin up the courage to take that hike again. And when I do, I'll be thinking of Richard Felger when I find that old *Salix bonplandiana* at the bottom of the box!

31

# HIKING THE GILA NATIONAL FOREST

Jane Spinti
with contributions from Betsy Kaido, Elroy Limmer, and Jeanne Martin

RICHARD FELGER MOVED to Silver City in 2007. He had visited our community a number of times previously, so by the time he settled here permanently, he had become acquainted with a group of hikers who belonged to the Gila Native Plant Society (GNPS). He soon began to join us on our weekly forays to places near and far in the Gila National Forest, where we would hike and identify plants along the trails.

Richard's interest in plants was infectious, but it didn't take long for us to realize that he was learning some of the plants along with us because many plants in the Gila National Forest do not occur in Arizona or northwestern Mexico, where he had done most of his botanical research. Yet he was a born teacher, and he loved to share what he knew, often with earthy allusions. When we left for a hike, he would ask us, "Are you dressed?" meaning did we have our magnifying loupes so we could look closely at various plant parts. Plants have males and females, or as Richard would slyly say, "Plants have sex," and he would encourage us to inspect their reproductive parts while he explained the intricacies of how they function. Douglas fir cones had mouse butts, and Emory oak leaves had furry butts, characterizations intended, perhaps, to help us remember what pine cones belonged to which pine species, and how to differentiate the oak trees.

It also didn't take long for us to discover that Richard was a true eccentric. He would often arrive for hikes at the last minute and sometimes rather flustered because he was sure he left behind something important, such as his phone, his notebook, or his wallet. Thus would begin a search through his backpack in the parking lot until the lost item was located or until he convinced himself that he really had lost or misplaced it. There was likely to be another search during our lunch break along the trail, which often revealed that the supposedly lost item had simply been buried deep in his crowded backpack. Sometimes this ordeal was repeated yet again when we returned to the parking lot, with Richard announcing that he had lost his car keys, followed by another backpack search. It was not unusual for Richard to ask for a ride home only to find his car keys as he arrived in his driveway.

Richard liked us to drive him to the trailhead so he could sit in the front passenger seat and botanize out the window on the way. When he saw something interesting he would call out, "Stop!" or "Slow down!" or "Go back!" even if we were traveling at sixty miles per hour. The entire caravan of cars would then stop, and everyone would get out so we could see the plants that he had spotted. Yet I never heard anyone complain about those stops. We respected Richard's knowledge of plants, and we all wanted to see whatever had caught his attention.

Richard frequently commented on the size of trees along the hiking trail. If he saw a particularly large specimen, he would ask one or two of us to stand next to it to provide scale while he directed someone to take a photo. In one memorable incident, he spotted a juniper that was full of pollen and asked a fellow hiker to shake the tree so that he could get a photograph of the pollen drifting off like thick smoke. Unfortunately, the pollen also covered the unsuspecting tree shaker.

Richard would often emphasize the importance of taking our time along the trail and being aware of what was around us. He could not understand why anyone would walk among plants and not wonder what species were present. He wanted to know each plant he encountered, whether it was common or uncommon, whether it was edible or not, and what other uses it might have had in the past. He made sure we noticed uncommon plants, such as blueberry plants (*Vaccinium* sp.) or unusual oaks (*Quercus* sp.), and would point out plants, plant parts, or unusual features by asking, "Do you see that?" "Do you see this unusual feature?" He was especially patient with new hikers who were trying to learn plants, making sure that they were included in the learning experience.

Richard himself was a leisurely hiker, taking in everything along the trail. He would stop frequently, and if he thought we had missed something important, he would call us back, often using the term "children": "Children, come and look here," he would say. It was his way of saying, "Please notice what is all around you." Sometimes we needed to wait for him to catch up. Hikes became so leisurely that most of us believed Richard could not maintain a fast pace. We were disabused of that belief one day when the sky slowly became overcast, and then it became obvious that it was going to rain. A simultaneous thunder and lightning strike jolted Richard into action, and we watched in amazement as he hurried back to the car at a pace most of us could not match. As he disappeared into the distance, we pulled on rain gear and followed after him.

Richard was very familiar with plants used by Native peoples. He often spoke of all the knowledge he had gained from friends and acquaintances among the Seri people of Mexico and the Apaches of Arizona and New Mexico. He knew which grains were collected for flour, which acorns were the most palatable (*Quercus emoryi*), and how to use mesquite flour. He frequently referred to one of his favorite grains, bulb panicum (*Zuloagaea bulbosa*), as Apache redgrass in an attempt to make us aware that this was a grass with a long history as a food source among Native peoples.

One of Richard's lifelong passions was identifying possible food sources that could feed a hungry world. He often asked us to think about how modern societies could fit plants to the land instead of vice versa: "What plants already grow here?" "How can we use the plants that are already here?" "What other plants will grow here?" Two grains he considered important were Apache redgrass and giant sacaton (*Sporobolus wrightii*). He began baking with both grains as well as with mesquite bean flour, and he encouraged us to try making foods with them. A memorable night at one of the monthly GNPS meetings included a table full of foods made by GNPS members. There were many dishes, all made from native plants, with enough food for all to share. On hikes he would taste, and encourage others to taste, native plants such as coffee berry (*Rhamnus*) or the acorns of Emory oak (*Quercus emoryi*), which, unlike most acorn species, taste sweet rather than bitter.

Richard was an inspiration, a teacher, and a friend. He was an inspiration for his unceasing pursuit of knowledge, for his desire to share what he had learned, and for his encouragement. For those who wanted his guidance, he gave it freely. For those of us who initially had only a casual interest in plants, he inspired us to look at them more closely, which motivated us to learn more about them. He encouraged us to use scientific names so we could better communicate with fellow botanists. He taught us to slow down and that becoming really aware of our surroundings made life much more interesting. I am sure that many of us think of him as we hike today, especially when we encounter a plant he taught us. We are grateful to have known him as someone who made a real difference in our lives.

32

# PRESSED FOR TIME

Sue Rutman

IN MY MIND'S EYE, I see Richard peering over his glasses at me before issuing a joke or snarky comment. Or I see him in his element, the Sonoran Desert, standing before a plant, writing notes that would eventually become part of a book. Richard, collecting handfuls of plants that would be added to an ever-growing bag of plants to be pressed. Me, looking at the long list of emails from him (up to fifty per day), some barely decipherable. Many emails telling me "this is the final" draft, only to find twenty more "finals" of the same chapter. Richard, asking me for the fifth time to send him a digital file he lost on one of his computers or dozens of thumb drives. Or telling me in a panic that his computer crashed (it didn't). Could I send him all my files? Or that I *must must must* do this or that. Oh, Richard.

I met Richard in the 1980s when he was self-employed, working in Tucson, and I was the endangered species botanist for the U.S. Fish and Wildlife Service in Phoenix. Richard and I would cross paths infrequently until I took a job as botanist at Organ Pipe Cactus National Monument in 1994. He quickly inserted himself into my job. Time and again I had to remind him that I didn't have time to explore and collect plants for him. Nevertheless, I'd occasionally get an email from him, ordering me to go somewhere and look for this or that species. I confess that I would usually fulfill his orders, but only on my time off.

In the late 1990s, Richard and I spent hours updating the monument's herbarium. Over time, we examined nearly two thousand specimens, with me at his elbow learning the finer points of plant identification. We annotated specimens as needed and updated nomenclature. During my tenure at Organ Pipe and for a few years after I retired in 2013, I collected about two thousand additional specimens, and Richard had to inspect every single one.

Richard was an excellent botanist, and his identifications were seldom wrong. Everyone makes mistakes, though, and on several occasions I questioned his identifications. He did not like to be caught misidentifying a plant. He generally denied having made a mistake, or he got flustered and made furtive excuses, and the correction was made quietly. I learned to correct his errors diplomatically. Sometimes I had to wait days or weeks for him to concede an error. The following email was Richard's explanation for misidentifying a plant that I eventually identified as *Eremocarya micrantha*. He sent me the following stream-of-consciousness email (original spelling and grammar retained): "oops i thot the calyx was really narrow—i just enlarged/ forgot to look closer—yr close up better than i thot—nah, not maritima, no bracts—multiple nutlets sorry too faasst—need nutlets to be sure—but prob just old scratchy J. angustifolia—does have what looks 4 nutlets—and maybe heterospermous R."

About 2010, a superintendent moved Organ Pipe's herbarium—some four thousand specimens—to the Western Archaeological and Conservation Center in Tucson, where the specimens reside in shrink-wrapped boxes in a dark room with limited access. Removing the herbarium from Organ Pipe was tragic. Richard was furious (so was I) and impolitely conveyed his feelings about the situation to more than one superintendent and Resource Management Division chief. His complaints fell on deaf ears, which added proof to his belief that most government workers were fools.

Richard spent a good deal of time at the herbarium (ARIZ) at the University of Arizona. His time there was sacred. He loved to work with specimens but was aggravated by people who interrupted him. He had work to do! No interruptions! No time to waste! Anyone who dared to interrupt was often met with withering rejection. On the other hand, he loved to teach and was good at it. If a botanist was struggling to identify a specimen, Richard would gladly lend a hand.

One of his treasures was a hand lens that was handmade by a friend from the wood of a large *Koeberlinia spinosa* from the center of Isla Tiburón, Sonora.

He would fly into a panic if he thought it was missing. Every now and then I was permitted to use it. I hope some lucky botanist inherited it.

After moving to Silver City, Richard continued to work on specimens at the U of A, even though getting to Tucson involved hours of travel and maybe an overnight stay. He disliked being away from home and avoided hotels, preferring to stay with friends. He sent the following note to me on November 26, 2019 (original spelling and grammar retained): "Life is good in silver city—if I can help it, I never leave home territory—i try not to drive in big city traffic and avoid big highways—when I have to go to tucson for a dr appt, I usually get a ride, and think about seeing friends etc—but end up running right back home as faaast as possible."

Richard was used to issuing orders and used to people jumping to fulfill them. People who worked with him a lot learned to set limits. Or not. Phil Jenkins, former assistant herbarium manager at the U of A, set boundaries when he told Richard, "I am not your slave." Fellow botanist Meg Quinn, who once spent a few hours listening to Richard ordering me around the herbarium, said to me, "I don't know how you stand it." I became selective about responding to his orders, so he stopped ordering me and started asking me for help.

When botanizing in the Ajo area, Richard sometimes stayed at our house. He was a gracious guest but couldn't resist ordering people around. One morning, Richard and I were pressing plants on the dining room table. Tim Tibbitts had just told Richard about a giant *Condalia globosa* along Kuakatch Wash. "You *must* take me," Richard said, oblivious to the fact that Tim was heading out the door to go to work.

Whether it was collecting plants in the field, working in the herbarium, or writing books, Richard was happiest when devoting his time to plants. I've heard people call him a genius, but I don't know what that means. I know he worked hard, almost constantly, and spent as little time as possible doing things that diverted his attention away from plants.

Richard spent seventy-some years getting to know the plants of the Sonoran Desert and adjoining areas, yet he was still learning about them until he died, with so much more to do, so much more to learn. No doubt he was asked the question most plant experts get asked: What will you do when you learn all the plants? The question might have sparked a rude reply. To Richard, the answer was obvious: There is so much to learn it is not possible to learn it all. He saw no end to his botanical work.

He loved going to places he'd visited many times or that he had never visited and seeing old plant friends or new acquaintances. He called me once to tell me he had completed a ten-mile hike in the Gila Mountains at his advanced age of eighty-plus years. He was thrilled to know he could still get out there and botanize.

Almost nothing could or did interrupt his quest for plants, but late in his life he was forced to deal with drug smugglers in "his" territory. They were the one threat to his personal safety that would send him into a frenzy and stop him in his tracks. One time after a collecting trip, we were driving southbound on Highway 85 through Organ Pipe and found ourselves following a black SUV with blacked-out windows and Sinaloa license plates. I pointed out the vehicle to Richard, telling him that Sinaloa was the source of much of the marijuana and opiates being smuggled north into the United States. Just as I was telling Richard about Sinaloa, the driver of the SUV stopped on the highway to point out an intersecting illegal smuggling road to his three passengers. When I told Richard what had just happened, he became agitated and yelled, "Pull back! Pull back!"

He eventually got so afraid of the drug cartels that he wouldn't go to Mexico anymore. He wouldn't go to Quitobaquito Springs either, even though it was one of his favorite places. In an email to me (November 26, 2019), he wrote, "I too am sickened by the border situation and creeping Nazism. I am glad we got our borderland work done before the current situation. I know I can never again go to Quitobaquito and retain a semblance of sanity."

Learning about plants was not Richard's end game. His botanical field trips, note taking, and plant collections were fun activities that provided material for his next book. And there was always a next book. He liked to write. He liked to teach people about plants through the written word. He also liked the thought of leaving a legacy.

His ability to juggle many projects was remarkable. He always had several papers and books and ethnobotanical projects in process, and always with one or more collaborators. While Richard and I were working on *The Flora of Southwestern Arizona*, he was working on two books with Sue Carnahan and José Jesús Sánchez-Escalante: *Oasis at the Desert's Edge: Flora of Cañón del Nacapule, Sonora, Mexico*, and *The Desert Edge: Flora of the Guaymas–Yaqui Region of Sonora, Mexico*, plus other publications with various collaborators.

He was all-inclusive in his search for and appreciation of collaborators, who not only gave him great pleasure and insight but also expanded his botanical family and increased his readership. While working on the Euphorbiaceae chapter of *The Flora of Southwestern Arizona*, Richard happened to meet a young botanist with no "credentials" who had wandered into the herbarium. It was Nathan C. Taylor, a student at Sul Ross University, who was studying *Euphorbia* species. Taylor was passionate about euphorbs and was quickly invited to be a co-author.

Richard never failed to give collaborators full credit for their contributions. His strong feelings about this issue were based on something that had happened to him early in his career. According to Richard, a well-known professor at the University of Arizona had taken full credit for a published paper that Richard had written. Decades later, Richard still fumed about it.

He started drafting "Grasses of Sonora" when widely respected grass experts John and Charlotte Reeder were getting old and frail. With some urgency, he met with them frequently at the U of A to record their knowledge. After they passed away in 2009, Richard continued working on the book with collaborators Pedro Garcillán, Jon Rebman, Jim Verrier, and Sue Carnahan. The book remains unfinished but would be a triumph if it were completed and published.

When he reached the age of eighty, Richard knew his time was short. He complained to me in an email, "In the years I have left I want to finalize way too many publications." "I'm like these herbarium specimens—I'm pressed for time" was one of his favorite time-worn jokes. In the months before he passed and while battling cancer, he finished or nearly finished three major publications with collaborators: *Plants and Animals in the Yoeme World*, *Field Guide to the Trees of the Gila Region of New Mexico*, and "Grasses of Sonora" (unpublished). Incredibly, he was lead author on three books that were published after his death.

Richard had an intense, complicated, and sometimes frustrating personality. He could be gruff and grumpy and funny and personable and gracious. He had a warm and generous side that too few people experienced. His affection for his friends was infrequently shown, but somehow it was clear to us that we were valued members of his family. Before he died, he sent his love to all, words that gave us a clue that he was nearing the end.

# 33

# YOURS TRULY, STUDENT OF BOTANY

Susan Davis Carnahan

ON OUR HANDS and knees, navigating a wet rock slope in Cañón del Nacapule and staring eye level at a mat of miniature sedges, Richard mutters, "Grab that! It looks important." I take a few shaky photos, pull a handful of the two-inch plants, and shove them in a plastic bag. They would turn out to be half-chaff sedge (*Cyperus subsquarrosus*, formerly *Lipocarpha micrantha*), new for the canyon flora. The next day, Richard finds the first record for Sonora of non-native jute (*Corchorus olitorius*, Malvaceae), also in Nacapule. That's about as wild as it got for my ten or so years working with Richard. No shoot-outs or short-runway takeoffs and landings with Ike Russell, no all-night drifting in a panga out of gas, no alligators, no collapsed lung on a dive, no roof-mounted plant press coming apart on the highway, no losing a wheel—a wheel!—from a pickup while driving. Not even a flat tire. Those adventures belonged to his earlier botany decades, which you can read about elsewhere in this volume.

Richard was eighty-one years old on his last visit to the Guaymas region, in October 2015. With his daypack slung over his right shoulder (the left one was often painful), his hand lens around his neck (he didn't feel properly dressed without it), and sturdy boots carrying him along the canyon bottom and upper trails, he was in his element. He even climbed the rope and plank ladder at the west end of the main canyon. "No photos," he grumbled as he made his way up. I took one anyway. He was playing hooky from the N-Gen Summit

Richard, age eighty-one, ascending the rope ladder in Cañón del Nacapule, October 2015. Courtesy of Susan Davis Carnahan.

conference in Guaymas, organized by his protégé and colleague Ben Wilder, who had bestowed on Richard a career recognition award, probably the sole reason Richard agreed to make the trip. Richard and I spent a day and a half in Nacapule together—my only experience in the field with him. And for the first time in my short botanical life, I was someone's field assistant. Clambering, yanking ("Wait, can't I get a photo first?" "Yeah, okay"), carrying, field pressing, and later also pressing Richard's small bag of ferns and other plant parts he had hoped would soften and uncurl with a bit of moisture (they didn't). But because it was Richard, I didn't mind.

My first contact with Richard was via email. In December 2011, I visited Cañón del Nacapule and took photo after photo of strange and unfamiliar plants. I sent a few to Tom Van Devender, who graciously helped with IDs but eventually suggested I send them to Richard, who of course had published a flora of the canyon in 1999. Richard responded promptly, identified every image I sent, and encouraged me to take and send more. ("Keep these—we might use them.") And we did—with Mexican botanist Jesús Sánchez-Escalante—for a revised flora of Nacapule in 2017, followed in 2023 by a flora of the entire region (Richard's lifelong focus): *The Desert Edge: Flora of the Guaymas–Yaqui Region of Sonora, Mexico*.

I didn't meet him in person until 2013, and it was the briefest of encounters. A beginning field botanist, I was at the University of Arizona Herbarium, struggling to identify a *Hibiscus*, and someone directed me to Richard, who was working through several folders of vouchers nearby. He pulled out his ever-present hand lens, peered at the stem, pronounced my plant to be *Hibiscus biseptus*, carefully explained the difference between the irregularly arranged stellate hairs of *biseptus* and the four-armed, neatly appressed hairs of *Hibiscus coulteri*, and returned to his folder stack. No small talk; I would later learn his Tucson visits were mainly about maximizing herbarium time.

At some point I told Richard I had been a copyeditor before taking up botany, and after testing me out on a few small documents, he decided he liked my skills (and my price, which was zero). Thus began a very fruitful working partnership and friendship. Richard loved to be copyedited, which is a rare quality in a writer, and a welcome one for an editor. He asked for and appreciated my suggestions, grammatical fixes, and rearrangement of sentences. "You are earning more chocolate and extra stars on yr chart," he would promise, though there wasn't really a chart, and the organic dark chocolate never materialized. We argued—gently—over his stylistic habit of leaving out details or juxtaposing big ideas without explanation. "No time to explain" was one excuse. "Leave a bit of mystery" was another. I would respond, "Mystery is great if it is done just right, but if not, it feels vague, and the reader just tunes out." He won some of those debates; I won others.

A hospital report from a 2018 emergency room visit listed Richard's occupation as *Botany Student, retired*. He loved this new identity and used "Yours truly, Student of Botany" in emails to me for the next eighteen months. He didn't have an actual job, of course, but he was far from retired, and he was clear about what he wanted to accomplish in these years: Finish flora projects, gather and edit the essays for this volume, promote crops for a drying world, lead local field trips, and do what he called show-and-tell at Native Plant Society meetings. Driving to Tucson for herbarium visits and doctors was necessary but *chronophageous*—time-eating—one of Richard's favorite neologisms. He worked at his computers every day that he could muster, despite eyesight issues, funding challenges, technical snafus, and health problems, including, eventually, the cancer that would end his life.

During my collaborative years with Richard, I spent a lot of time in San Carlos, Sonora. My husband, Curtis, and I kept a sailboat there, which gave us a place to stay as well as something for Curtis to work on while I roamed about with a camera and plant press. Richard eagerly awaited my new photos and specimen lists, helped me with plant names, gave me homework ("check *Croton* subfamily for color of sap—esp. *C. magdalenae*"), and rolled his eyes—I could tell—whenever I found a new urban weed. Richard had already collected his share of weeds, including a veritable haul of one-off exotics at the Guaymas airport in 1985, but at this point in the flora, each new weed meant writing new species accounts (and sometimes family and genus treatments) and revising keys, tables, and flora counts. So, he urged, "if you see a new weed at San Carlos, pull it up and put in trash unless you cannot resist temptation."

Richard always worried I would have trouble bringing certain specimens across the U.S. border. "Call every morning-glory *Ipomoea arborescens* [the tree species]" so the USDA inspectors wouldn't seize them as agricultural weeds. "Label *Cuscuta* [dodder] as the plant it's growing on." "Don't use the name *Euphorbia* or they'll think it's a commercial cactus lookalike." And so on. I followed some of his advice, but not all of it, and I didn't have any issues. "Do you know Richard Felger?" inspectors used to ask me. Yes, I always said, unsure if I should have said no. In summer 2020, when we were finalizing the *Desert Edge* text, Richard wondered, "Should USDA ag inspectors be acknowledged—they have a tough job and are prob underpaid and never given credit?" I responded that they didn't cause too many problems, but they didn't exactly *help*. The USDA did not make it into the book.

Eelgrass (*Zostera marina*) and other sea grasses were an obsession for Richard that resurfaced every time I went to San Carlos. He had seen *Zostera* there in 1977 and again in 2000 but never collected a specimen for the flora. "Look for sea grasses on the open beach." "Look for *Zostera* in the beach drift." "Check beach drift at any sandy beach—G region—esp. near Pilar, and Miramar—for sea grasses . . . do not expect seagrasses on open shores north of san carlos." I never found any, but in April 2019 I wrote to Richard:

> *Curtis had a dream last night that we found seagrass! The tide had gone out and it was anchored in the sand.—Sue*

> He replied: *Good enough! Yoemem / Comcaac too—say dreams are real— . . . Curtis (pers. comm.)—R*

In my early days, I treated botany hikes like treasure hunts or scavenger hunts. Wasn't it everyone's quest to find rare plants? Richard congratulated me on uncommon finds like *Helicteres baruensis* or *Salvia palmetorum*, but most of the time what he wanted to know—about the new canyon I explored or an anchorage we visited—was how many *chalates* (*Ficus insipida*) and *nacapules* (*Ficus pertusa*) I saw, were there any *Washingtonia* palms, and what were the dominant succulents and woody plants and grasses. The vegetative community, the regulars, the backbone of the flora—those were his way of interpreting the world. In Nacapule in 2015, Richard greeted the palms and fig trees like old friends. He nodded at the common desert broom (*Baccharis sarothroides*) and hopbush (*Dodonaea viscosa*), admired the fast-growing Goodding willow (*Salix gooddingii*) that had appeared since his last visit, and remarked on the weedy *Waltheria indica* that had spread throughout the canyon. It was not all about the surprises, but instead about history, continuity, trajectory.

As serious as he was about botany, Richard loved to include a good story in a flora. For example, this observation of *Plumeria rubra* (Apocynaceae):

> Cañón La Pintada, Sierra Libre, 31 Dec 1983, T.R. Van Devender, with R.K. Van Devender, observation (MABA, SEINet; email from Felger, 29 Jul 2016: "*Hi Tom, Do you have any more information for this record?*" On 29 Jul 2016, Tom wrote: "*I remember it clearly. Leafless stems, multi-stemmed shrub, 1.8 m tall, in a rock wall in the upper steep part of the trail opposite the upper part of the paintings. On our honeymoon.*")

Or my collection of *Ipomoea arborescens* (Convolvulaceae):

> Mex Hwy 15, about 55 km N of Guaymas and 15 km S of turnoff for La Pintada, 12+ trees on rocky hillside E of highway; leafless flowering tree 3.5 m tall, 28 Jan 2018, *Carnahan 2853* (ARIZ, USON; Sue writes: "*We have been passing those palo santos for years but Curtis managed to drive right into the construction site [where we woke up the guy who guards the equipment] so I could crawl through the fence for a specimen.*" Richard writes: "*There used to be an arroyo there; that is where I first saw palo santo trees, and in flower, on my first trip to Álamos, when I was 15 years old, with my high school biology teacher Nancy Thomas, her boyfriend Peter Neely, and UCLA researchers from Prof. Raymond Cowles' lab.*")

Richard also liked to turn nonbotanists into collectors. One day as I scrolled through online specimen records from Sonora, I stumbled across an unusual name, but one that I recognized: Nanao Sakaki. How many people could there be with that name? The Nanao Sakaki I knew was an itinerant Japanese poet whom I met in the 1990s when he was a guest of Gary Snyder in Northern California. Apparently Richard knew him too. Nanao and his partner Nancy accompanied Richard to the Sierra Pinacate in April 1970, where Richard collected nearly forty specimens, but Nanao is named as collector of at least four of them. The vouchers all have five-digit numbers, obviously Richard's numbers: "*Senegalia greggii* (Fabaceae): Mexico, Sonora, Puerto Peñasco, near Emelia Tank, Pinacate region; 20 Apr 1970, *N. Sakaki 19729* (ARIZ), with Nancy Tucker, R. S. Felger."

Although Richard was often an early adopter of new plant names and nomenclatures, he was more of a lumper than a splitter of species. He considered the desert lavenders *Condea albida* and *C. emoryi* a single variable entity. He treated the bristlegrasses *Setaria leucopila* and *S. macrostachya* as a species complex, after grass specialist John Reeder identified collections from the same place as *leucopila* in a dry year and *macrostachya* in a wet one. "I am a bit leery of the splits," he wrote in 2020 about a proposed division of *Talinum* species. In his decades of field experience, Richard had encountered the same species over and over again, in shade or sun, during a drought or after a hurricane, on a north slope or a south-facing cliff, in a cave or on the beach. He knew firsthand that an individual plant might be hairy or glabrous, glandular or not, huge with extra-large leaves or flowering at two centimeters tall, regardless of how it was described in the literature. "Plants don't read the books," he was fond of saying, and "That's what grasses will do," echoing the agrostologist Agnes Chase. The harshest criticism I heard him level at a botanical treatment was "Taxonomy is fun until it isn't."

Even Richard's namesake plants were not immune to his lumping tendencies. In *The Flora of Southwestern Arizona*, he reduced *Chamaesaracha felgeri*, a nightshade family species described as endemic to dunes of Yuma County, Arizona, to a synonym of *C. aridus*, believing that the large leaves of *felgeri* were due to abundant rainfall rather than genetics. That's not to say he wasn't proud of "his" plants: *Isocoma felgeri*, endemic to the Bahía de Kino area on the Sonoran coast; *Verbesina felgeri*, best known in Cañón del Nacapule, where Richard posed reluctantly with it in 2015; and especially *Agave felgeri*, a miniature century plant endemic to the Guaymas region and described by Richard's hero, Howard Scott Gentry, in 1972. He liked saying that plants named for him were usually scratchy or poky.

I received many emails a day from Richard, sent one after another as new questions and thoughts popped into his head. They were typed like text messages: short and telegraphic, with little regard for spelling or punctuation: "i am trying to do clean up of a million emails. maybe I added to confuesment—gaz etc Ráhum Pótam—accents (but NO accents in Yaqui language). maybe I already sent this in the late Pleistocene." They were also playful. "Pax et Panicum," he signed off when worried about being too slow on a project. "Good—will wait for your next missile (missive)." That was the poet in him—he loved the layers of meaning in words and especially the gratuitous word play of typos. "I am taming the Marcus Jones listings and then onto the rest of my G [Guaymas] homeworm. (thats a good typo)." He referred to Salero Ranch, where I live, as Saltshaker Ranch after I told him *salero* meant "salt dish." He also called it Salt Boat Ranch. During editing of the *Desert Edge* manuscript, we were asked to remove uses of "see" and "see also" from text citations, such as "(see also Felger et al. 2011)." Richard was annoyed, thought it petty and ridiculous, and was getting ready to just say no when Curtis suggested, jokingly, "What about *look at* instead of *see*?" Richard loved that, forgot about his annoyance, and for the next three months peppered his emails with "look at."

*gimme a little time to 'look at'*
*always doubt authority (look at)*
*after more "look at"—I think the unbold shd be bold*
*but you may have a better opinion—and we did this a while ago (below, look at)*
*"look at" (C. Smith, pers. comm.) is our best one*

Until the last several weeks of Richard's life, he and I rarely spoke on the phone. In fact, I spent so many more hours emailing than talking to him that this near-daily exchange was the voice I knew best and miss the most. Reading through all the old messages, I can hear him again:

*everything the same here except there is more of it*
*super photos—glad it is too dry to find anything new -*
*there aint no time i aint busy but i got time for this . . . doing it now*
*glad to see your edits are getting fewer and shorter -R*

# 34

# THE RIGHT QUESTION

Benjamin T. Wilder

SOMETIMES IT ALL STARTS with a question.

"What plants occur on Isla Tiburón?" I innocently asked Dr. Felger.

"We don't really know. Why don't we go there, fill in some gaps, do the flora, and write a book," he enthusiastically replied.

This was before we figured out that his first date back in 1940s Los Angeles was with my great-aunt Esther Capin, and before we realized that we both are descendants from Jewish merchants who established businesses along the U.S.-Mexico border in the early twentieth century. It was certainly before we shared island-hopping adventures, hours upon hours at the herbarium, even more hours with back and forth editing and consulting, and a friendship that opened the world to me.

Back to 2005, the University of Arizona Herbarium, and my twenty-year-old self. I could barely offer a response to such a proposition. In my mind I was thinking, "This guy is crazy." I gathered myself and said, "I have no idea how to do a flora or where to even begin." He succinctly said, "This is how you learn."

I left the herbarium scratching my head, unclear what to make of this whole thing. That night I wrote an email to Ray Turner, USGS desert plant ecologist and grandfather to Seth Turner, my best friend since first grade, and extended grandfather to me. I explained the situation to Ray and asked what I should make of this and who is this Dr. Felger guy anyway? The response came quickly the next day. "This is the opportunity of a lifetime." That was all I needed to hear, and as in all things, Ray was right.

In December 2005 Richard sent me off on my first trip to Isla Tiburón, the homeland of the Comcaac (Seri people), to begin our island flora project, for which I had the company of my good friend Ed Gilbert. Richard's friends and Seri linguists, Cathy and Steve Marlett, had recommended that I contact Humberto Romero Morales, who lives in the Seri village of Punta Chueca and has great knowledge and passion for working with plants. Richard had given me a copy of his and Becky Moser's out-of-print *People of the Desert and Sea* to give to Humberto. Loaded with plant presses made of cardboard sheets tied between two pegboards with parachute cord, and more *Arizona Daily Wildcat* newspapers (then the exact size of an herbarium sheet) than seemed reasonable, Ed and I set off.

We made our way from Bahía de Kino to Punta Chueca on a road that was a decade away from being paved, as a flat tire beyond repair en route can attest. We were directed to Humberto's house, and he was there. I introduced myself and Ed, told him of my goal to catalog the flora of Isla Tiburón, and mentioned that Cathy and Steve had recommended I work with him. He lit up when I handed the Seri ethnobotany book to him and instantly started looking through it, relating information to us as he saw plants he knew. He then said, "Okay, let's go to the island," and was instantly ready, as I, in shock, assembled my loads of field gear, bags for collecting, and plant presses.

I am still amazed at how quickly we were whisked across the Canal del Infiernillo to the island, then into one of the SUVs used in bighorn sheep hunting on the island, and up to the base of the towering Sierra Kunkaak. We dismounted from the car and tried to keep up with Humberto as he set a blistering pace through the dense desert scrub. Ed and I, with two huge bags, started collecting a sample of every plant in sight. We were following a wash near a waterhole where we wanted to collect, but Humberto really wanted to show us some of the higher-elevation vegetation. Along the path Ed saw a smoothed rock and picked it up. Humberto said, "*Es de personas antes*." It was an ancient grinding stone. Past steep hillsides, we emerged on a beautiful view of an arroyo and the Sierra. It was quite green compared to the rest of the island's vegetation. Humberto stopped, and I could tell we had reached our destination. He sang a song in Cmiique Iitom (the language of the Comcaac) for the vegetation. The new journey I was on that connected to a place and people so much larger than anything I had previously known

Humberto Romero explaining the spiritual importance of the cactus *hasahcapjö* (senita, *Lophocereus schottii*), Isla Tiburón, May 2006. Photo by Ben Wilder, originally published as figure 3.57 *Lophocereus schottii* from *Plant Life of a Desert Archipelago* by Richard Stephen Felger and Benjamin Theodore Wilder, in collaboration with Humberto Romero Morales © 2012 The Arizona Board of Regents. Reprinted with permission of the University of Arizona Press.

began to open around me. He finished and said, "Okay, where do you want to go?"

Ed and I came back from that trip with over one hundred plant specimens for herbarium collections, several of which were new records for the island. I think I passed "the test," and Richard was all in. That following May he joined me for his first trip back to the Seri region and the island since 1983, which had been the last of his dozens of trips that spanned from his dissertation to the production of the ethnobotany book. We made several more trips in the following years. Each trip built on the last, and the scope of the project grew. If we are doing the flora of Tiburón, he said, we really should include the adjacent islands too. And Isla San Pedro Nolasco, off the coast of Sonora near San Carlos, and we can't not include the most isolated island, Isla San Pedro Mártir.

In April 2007 we set off to Mártir, sixty kilometers south-southeast of Bahía de Kino, with Richard's longtime dear friend Exequiel Ezcurra. I was a fly on the wall for the hour and a half boat ride to the island, dazzled by the far-ranging conversation between them, years after their last time together. They shared their fascination and love for Argentine deserts and plants, and Richard noted that the Monte Desert of Argentina does not have the climatic extremes the Sonoran Desert does. He pointed out that there are a lot of horticultural plants, especially from New Zealand, that do well in the Monte that can't survive in the Sonoran Desert. As we passed by some shrimp trawlers, Richard and Exequiel commiserated about the subsidies paid by the Mexican government to the fishing industry, and how that artificial support contributed to the continued depletion of the Gulf fishery. Exequiel excitedly told Richard about one of his current projects, which focused on the travels and legacy of Edward Nelson and Edward Goldman's partnership as they explored the natural history of Baja California and elsewhere in Mexico at the beginning of the twentieth century. Exequiel had uncovered a trove of Nelson's unpublished glass plate photos and associated lecture notes at the Smithsonian Institution, which Nelson had used for the public presentations he once delivered. One of the photos was an enigmatic scene that Exequiel suspected might be from Isla San Pedro Mártir, and his desire to match the scene in part drove his participation on this trip. (Though he was unable to locate the scene, which was actually from bird islands off the coast of Guerrero, later that year Exequiel did publish a beautiful book, *On Horseback Through the Deserts of Lower California*, with Nelson's original photos and notes.) While they were discussing book projects, Richard asked Exequiel about scientists who had studied the Sierra Pinacate for his forthcoming book with Bill Broyles, *Dry Borders* (also published later that year), which prompted Exequiel to launch into stories of early Pinacate explorers.

I sat in awe at the worlds of knowledge these two joyfully shared as we passed over the Gulf waters on a calm day toward San Pedro Mártir. Little did I know that I was just two years away from joining Exequiel at the University of California, Riverside, for my PhD as I continued to study the historical biogeography of these very islands. While I may not have been aware of what was afoot, Richard sure was.

Another trip to Isla San Pedro Mártir with Richard in December 2007 is also etched into my memory, though for different reasons. The Mexican Commission of Natural Protected Areas (CONANP) had just led a heroic

Exequiel Ezcurra and Richard deep in discussion among the dwarf cardons of Isla San Pedro Mártir, April 2007. Photo by Ben Wilder.

project that eradicated the invasive black rats from the island, and we were going there to establish a set of permanent plots on top of the island to track the demography of the island's forest of cardón cacti (*Pachycereus pringlei*). (I continue to monitor these plots every five years, which document one of the densest cactus populations in the world.)

On December 5 we departed from Bahía de Kino in the *Orca*, CONANP's panga (a twenty-five-foot open boat) with CONANP's boatman Miguel Durazo at the helm. In addition to Miguel, Richard, and me, our group included Andrea Galindo, who was a fellow at the Prescott College Field Station in Kino, and two Kino fishermen, Rafael and Bonito, to assist Miguel. We arrived at San Pedro Mártir at 1 p.m. and spent the rest of the day setting up camp and hauling our gear to the island's summit.

The next two days Richard, Andrea, Miguel, Rafael, and I established our plots, and by late afternoon of December 7, it was time to head back to the mainland. Looking out at the water, I saw that the wind, which had been strong earlier, had died down, and conditions looked calm. We departed in the *Orca* as rays of golden light broke through low clouds, and the sun set behind the Baja California peninsula. I expected to soon be back in Kino enjoying a seafood dinner. But after about an hour and a half, Miguel slowed the boat and said that we still had a long way to reach the mainland. We could keep going, but it would be better to head for a protected bay on the southeast coast of Tiburón and get into Kino in the morning. We listened to our captain, and he changed direction.

At 7:07 p.m. the engine stopped. An ominous sound followed by a discomforting statement from Miguel: "*No hay más gasolina*." There was a leak, and all the gas had drained into the ocean. To make the situation worse, the radio was not functioning, and my Mexican cell phone received no coverage at sea level.

The recent death of acclaimed scientist Gary Polis and four companions, whose boat had capsized across the Gulf in Bahía de los Ángeles, haunted me. If we drifted to the south, our fate might be the same. I quickly tried to push this out of my consciousness with deep breaths.

Because Miguel had changed course, we were in sight of Isla Dátil, just south of Tiburón. I got out my GPS and marked our location after the gas ran out. With the "tracks" setting I was able to gauge our movement toward or away from Dátil. To our great relief, after fifteen minutes it showed progress toward the island. But we had a long way to go.

More vain attempts with the radio, flashing SOS with lights, and firing a flare, which cast an eerie red glare against our faces, an image seared into my memory. Richard had a large tarp, and he, Andrea, and I held it up to create a forward sail, which caught a good deal of wind. Miguel got out two oars and began paddling toward the island. Dátil was getting closer and closer. Checking the GPS, I confirmed this. The deep breaths I had used earlier to calm myself were replaced with a sense that we would gain the island.

It was dark, and we were cold and wet. We settled in for a long night, not easy with six people and three hundred pounds of jumbled gear in a small boat. We tried to divert our minds from focusing on reality, and we each found a quiet space that allowed us a bit of rest. Lying in a disabled water-soaked panga, desperately cold, next to my good friend Andrea and my mentor Richard, with the entire universe of stars spread above and a dancing sea of phosphorescent lights below, I had to pause and take in the beauty, breathing it in through the cold and fear. After five hours of paddling and bailing, we set anchor about twenty meters from the high cliffs of Dátil, and it was time for sleep, if you can call it that.

About 5 a.m. we were awakened to the sound of waves crashing. The anchor had slipped, and we were just meters away from the steep sea cliffs of Dátil. Miguel was paddling the oars backward with all his might. He gradually regained control and pushed us away from the cliffs and through the swirling currents.

As the sun crossed the horizon, we saw that we were on the west side of Dátil, and we could see Islas Cholludo and Tiburón to the north. We spotted a beach with an accessible route to the island's crest. I needed to get up there and make a call with the cell phone. Miguel and I jumped into the surf and ran onto the beach and up the loose rock. Solid ground, even if a sixty-degree angle, was a true relief. We made it to the top, and I anxiously turned on the phone. The seconds passed . . . three bars! Miguel called his wife, María Jesús, and delivered the message of our location. She said she was headed over to alert Tadeo (Tad Pfister), the Prescott College Station manager and good friend of us all.

We paddled through strong waves around the north end of the island and finally gained safety on a gentle beach on the northeast side with minimal wave action. Fourteen hours since the engine quit. Time to wait and get dry. Then it was time for breakfast. The menu consisted of fish caught at least a day ago, fried in oil over a wood flame, with seawater-soaked tortillas from Anita

Street Market in Tucson. Soon Tad and Cosmé Becerra arrived in the Prescott College super panga, and we accorded them the role of heroes. We called this place Rescue Beach and, not forgetting our botanical mission, collected *Lycium brevipes*, *Wilder 07–608*. We headed for Kino with the *Orca*, refueled by Tad, alongside. We had all acquired an intense respect for travel in the Gulf and were on the good side of a great deal of luck. Richard may have best summed up our feelings after this whole experience: "I feel like I have been in a great fight. I won, but I am wounded."[1]

After a couple more years, many additional island trips, herbarium weekends, endless drafts, and countless emails, Richard, Humberto, and I did publish that book Richard first spoke of, *Plant Life of a Desert Archipelago*. Thankfully our friendship and partnership did not stop there. Throughout my doctoral work and subsequent research on the biogeography of the Gulf islands, Richard was by my side guiding me through questions, writing, and making sure the process stayed fun. Then I in turn began helping him publish some of his important but unfinished works. The last time we spoke, he knew his end was near, even if I didn't. He implored me, "NO MORE CHANGES!" to his books on the regional flora of Guaymas and his Yoeme (Yaqui) ethnobiology. "They are done."

While Richard left us far sooner than I ever would have expected, I don't feel far removed from his presence. I frequently channel his guidance when identifying a plant, writing a paper, or working in the field. I can feel him pushing me to go that additional step and make sure to get it correct. He taught me how to be a scientist, a writer, a field botanist and ecologist, an interviewer, a grants getter, a friend, a teacher, an explorer. He taught me that life and work are to be fun. Purposeful and meaningful, but *fun*.

I still reflect with wonder on the openness with which he greeted me on our first meeting and the amount of time, knowledge, and patience he invested in me in the following years. The right question did indeed lead to the opportunity of a lifetime. His gifts will last my lifetime and gave me a mentor that only the luckiest of us find.

---

1. The remembrance of being lost at sea is adapted from *Plant Life of a Desert Archipelago*, by Richard Stephen Felger and Benjamin Theodore Wilder, in collaboration with Humberto Romero Morales © 2012 The Arizona Board of Regents. Reprinted with permission of the University of Arizona Press.

# PART III

# ABOUT RICHARD FELGER

Bill Broyles

THERE YOU HAVE it. A first telling of Richard Felger's magnificent, wonderful, crazy-good life, life as he made it and chose to live it. His was a career founded on methodical observation and peer-reviewed fact, but his life was fueled by blazing curiosity, stirred by contradictions and connections, and sustained by steadfast hope. Plants were the very air he breathed. He seldom spoke of his personal or family life, leaving us with unopened chapters of his story and few glimpses of his private-most thoughts. Stoic and laser focused on his mission of the day, he was relentless and seldom off course. His work was his life, and he lived to work.

Biologically, he was a male of European descent, standing about five feet ten and weighing 175 pounds. Hair: brown and short, not needing a comb. He was clean shaven, except for a well-trimmed beard in the mid-1970s. Eyes: green. Right-handed. His credentials included a driver's license, doctorate degree, passport, voter's card, and an ever-present pocket notebook. At home he preferred to "graze" instead of dish up a whole meal three set times a day. He was a self-sufficient cook—if he remembered to bring groceries—an adaptable eater if you served the food (the tastier the better), and a voracious reader but a fitful sleeper, often reading until after midnight and logging on to his email before dawn. His preferred exercise was swimming, but walking sufficed, and at times he could be found in a gym lifting modest weights or doing aerobics.

Richard Felger at Campo Dólar, a traditional camp of the Comcaac on the Sonoran coast, December 1981. The granitic rocks in his hands are probably abandoned manos—hand stones used by Comcaac for grinding seeds. Photo by Irving Davis, courtesy of Cathy M. Marlett.

When driving alone, he listened to National Public Radio, at high volume. At times in his life he kept pet dogs and cats, juvenile alligators (Alley Oop, JoJo, and Swallow), a blue-fronted Amazon parrot named Tarzan, and other menagerie. His kinkajou "got banished just b/c he shredded the living room curtains and liked to poop between the couch cushions that [a guest] discovered after plopping down."[1]

When Richard went to grade school, he asked the chauffeur to drop him off several blocks from the school grounds so the other kids wouldn't tease him about being rich, and he memorized his route not by street signs but by the landmark plants growing along the way—teaching himself their Latin names, not the common nicknames used by his family's gardener.

He devoured botanical journals, *The New Yorker*, *Smithsonian, The Economist*, *New Scientist*, and *The New York Times*. He amassed a considerable research library, as well as a significant collection of poetry books centered on the Beats of the 1950s and '60s. On his walls he hung an impressive collection of authentic Yaqui (Yoeme) ceremonial masks and folk-art paintings by John Brandi and "Mac" (Mary Alice Cox) Schweitzer. His homes and patios burst with potted cuttings or stock seeded from his trips, and a grand assortment of arid and semitropical plants thrived in his yard. A *Welwitschia* from Namibia might be potted next to an orchid from South America and be shaded by a velvet mesquite native to Arizona.

On trips he would "ride shotgun" in the front passenger seat and scratch out his observations of mileages, species seen, and comments on the terrain, ecology, persons met, and weather into a pocket notebook. He could identify hundreds of species at sixty miles an hour, had taught himself to print legibly on even the bumpiest of roads, and might ride for hours or sit in camp without a word about any topic except plants. Then he'd pop out with, "I should have married her." "Who, Richard?" "My French girlfriend. She was killed in an elevator accident. So tragic." And then he'd resume his work at hand.

Or in a phone conversation, he might mention reading reporter Seymour Hirsch's dispatches in *The New Yorker* about war in the Middle East, and then he'd say, "I don't know where to go, Sweden or Australia? I'm afraid they'll come after me." During the war in Vietnam, he toyed with moving to Perth. Although he wasn't overtly Jewish, he deeply remembered family stories from

1. Richard Felger email to Bill Broyles, April 21, 2017.

Richard's house at Hacienda del Sol, the patio festooned with his vast collection of succulent plants from around the world, June 1981. Richard Felger archive.

his childhood about the pogroms of mass resettlement, imprisonment, and murder, especially of nineteenth-century Jews in Russia, as well as Kristallnacht in 1938 during the Nazi rule of Germany. His immediate ancestors—the Felger, Pincus, Lewkowitz, and Bernstein families—were all threatened by the pogroms, and many had fled Ukraine, most to the United States. Occasionally he'd voice a doomsday dread of a Fourth Reich.

In college he had been a sharp dresser, but later he was so unstylish that he had his own style. Often he relied on female friends to shop for his clothes, and he asked others where to buy hiking boots or tents. He didn't know, and he didn't care that he didn't know. During his later years, in the field or office his uniform was a loose long-sleeved work shirt with two button pockets and maybe the sleeves rolled up past his wrists, long durable slacks—but sometimes denims—and a wide-brimmed hat. A botanist's hand lens always—always—hung around his neck so he'd not appear "naked," and in one hand he carried a plastic or cloth bag for clipped plant parts, which he later pressed. He wore no jewelry, had no tattoos, and seldom seemed to sweat. His demeanor was calm, unflappable it seemed, until he spotted a favorite plant or a new species for the locale. Then he'd rave uncontrollably with excitement:

"Stop, stop, stop. Hot shit! A *Mirabilis tenuiloba*!" Or he'd spot a teacup-size *Fagonia laevis* in bloom and stoop down to cradle it in his hand as one would a jewel, marveling at the beauty and wonder of Life itself.

While packing for a camping trip to the Cabeza Prieta National Wildlife Refuge west of Ajo, one passenger dithered about which plant book to pack. Matter-of-factly and with no sign of hubris, Richard volunteered, "Oh, you don't need a plant book. You've got me." His zeal for plants was contagious.

On what were called "hikes," we knew that Richard was really searching for plants to identify, bag, and press for an herbarium. Plants were fodder for his next book or paper. Though he enjoyed vistas and sunsets, they were incidental to his main mission. But whatever the heat or wind or troublesome terrain, he never complained, not about his sore feet or aching shoulder, preferring instead to extol tiny belly flowers or measure cholla spine lengths with his just-out-of-the-box digital micrometer. He was quick to adopt the newest technologies—computers, big screens, color screens, laptops, thumb drives, and pocket-size digital cameras—and put them to work for his projects.

After a day collecting plant specimens, he would forgo supper until his plants were pressed and labeled, like a cowboy who made sure his horse was fed and watered before he filled his own plate and cup. Then before sunrise, he could be heard quietly unstrapping his plant presses to replace damp blotters with dry, making certain that his specimens were "comfortable." On a trip, this would be done morning and evening, as regularly as a monk going to prayer. His only complaint about camping recalled a time on Isla Tiburón when mosquitoes and *jejenes* kept him and friends awake all night. Although he didn't look the part of macho man or a commando prepared for any discomfort, no one was more at home in a raw camp than he was—after all, he had camped many nights with austere bush pilot Ike Russell and the durable Seris, sometimes without so much as a blanket or sufficient water, let alone comfortable camping gear. Yet he faithfully completed his field notes before turning to sleep.

His publications list is extensive, reflecting nearly seventy years of fieldwork, study, and writing. Friends describe him as purposeful, focused, intense; he never was one to "kill time" or "hang out." In groups, he might offer to shake your hand, but he was not a hugger. Or he could stand next to an old friend and not even say hello to them—you might think, "That's strange," until learning that the two of them had exchanged a dozen emails before breakfast that morning, or he was staying at their home while in town. He was reserved, even a bit shy or detached like an observer, not the life of the party, but after a meeting or fiesta, he probably would be the one you remembered for his attentiveness, insightfulness, or graciousness. And your name would be in his notebook. Even in his early books he acknowledged scores of people who helped him gather information or photographs. Villagers and fishermen were credited alongside professors and funders. His acknowledgment in *Flora of the Gran Desierto* runs two full pages and reads like a yearbook roster. His plant labels usually recorded the names of anyone who was on the trip, whether they recognized the specimen or not.

## ANCESTRY

So who was this Richard Felger, the botanist whom the Seri Indians of Mexico called "The Guy Who Cuts the Tops Off Plants?"[2] And who were his ancestors? Richard's father, Louis, was born in Cincinnati, Ohio, on September 23, 1892, and died February 24, 1971, in Beverly Hills, California. He was the only child of Solomon Felger and Eva Bernstein Felger. He earned his MD from the University of Southern California and license to practice medicine in 1914, before going to work at the Los Angeles County Hospital. From 1916 to 1919 he served as a doctor in the army medical department and was on duty with Allied forces during the Meuse-Argonne Offensive of 1918, the last battle of WWI. By the 1930s census, he was well established, and he gave his occupation as surgeon and general practice, though at other times he identified himself as a personal physician, in essence a doctor to the Hollywood stars and doing house calls in Beverly Hills and Malibu. Louis preferred to be called Louie, even professionally, and his obituary cites him as Dr. Louie Felger.

Richard's mother, Sadie Dorothy Pincus, was born September 17, 1904, in Butte, Montana. We know little of her life there, but according to Richard, "My mother's family, from Butte, Montana, spent a lot of time in San Francisco and Seattle. You know how bad the winters were in Butte if they went

2. Super-sleuth Sue Rutman graciously combed the internet for Richard Felger's family history, and Richard's cousin Karen Glikman generously provided family records and information. We consulted Felger's website, his University of Arizona Herbarium website, vitae, and bibliographies in our files, as well as undated conversations with his friends.

Louie and Richard Felger dressed for a formal event, 1950s. Despite Richard's general aversion to formality, we speculate that this special occasion was his first marriage, which took place on July 21, 1956. Courtesy of Silke Schneider.

to Seattle for part of the winter."[3] Too, the Pincus family had connections with Jewish families in Tientsin, China, and San Francisco and Seattle were gateway ports between the United States and China. Scattered clues indicate that Sadie's parents had known Louie's parents in Butte. Her father eventually

3. Richard Felger email to Karen Glikman, June 27, 2012.

owned land and property in Southern California, where the Felgers had settled, and the two families stayed in touch. Sadie attended the University of California, Berkeley—she's certainly in the school's 1924 and 1925 yearbooks and likely others, and she enjoyed the attention of several suitors. Richard remarked, "Sadie once had a little box full of fraternity pins and a few engagement rings," and much later, he met the son or grandson of one of her former "beaus."[4]

On a visit to Los Angeles with her mother in 1927, the Pincuses' smart and winsome coed met the Felgers' handsome young doctor. They married within a year at the Mark Hopkins Hotel in San Francisco. Their honeymoon trip to Hawai'i by steamer made the local newspapers, and in Honolulu they stayed in the Royal Hawaiian Hotel. Upon their return to Los Angeles, they resided at the Ambassador Hotel.[5] They had two sons, Dan and Richard. In 1974, three years after Louie died, Sadie married attorney Martin Gang.[6] Gang's will amply provided for his four children from a previous marriage and for Sadie, who lived her last years in Beverly Hills with a care staff. Sadie died June 1, 2009, at age 104. In part, her obituary reads, "Sadie lived her life with elegance, grace, integrity, and love."[7] We've not found any comprehensive accounts by Richard about his family, but he did submit snippets to a cousin.

According to Richard, "As you know, I was born in 1934. My father was 41 years old, and my mother, Sadie Pincus Felger, 29 years old. At that time they

4. Felger email to Glikman, June 27, 2012.

5. "Waded Through Rice," *Montana Standard* [Butte], September 13, 1928. The wedding was held in San Francisco because Sadie's sister Frieda (Mrs. Clarence Levy) lived there.

6. Martin Gang (March 12, 1901—January 29, 1998) was a partner in the Los Angeles law firm of Gang, Tyre, Ramer and Brown, and a founder, chairman, and president of American Jewish Committee's Los Angeles chapter. He represented such Hollywood stars as George Burns, Bob Hope, Olivia de Havilland, Myrna Loy, Lucille Ball, and Frank Sinatra, and provided legal aid to victims of the 1940s and 1950s Hollywood anticommunist blacklist. Additionally, he rescued members of his extended family living in Vienna, Austria, from Hitler's Holocaust. Gang promoted good relations between Jews and Christians and established the Martin Gang Institute for Intergroup Relations Training. *Los Angeles Times* obituary, February 4, 1998; *New York Times* obituary, February 1, 1998; Wikipedia, "Martin Gang," last edited on December 29, 2025, https://en.wikipedia.org/wiki/Martin_Gang/.

7. "Sadie Felger Gang," obituary, *Los Angeles Times*, June 3, 2009.

Richard's childhood home at 117 S. Las Palmas, Los Angeles, California, mid-1950s. It was designed by noted architect Heth Wharton and built in 1936. With nearly five thousand square feet, it had four bedrooms and five baths. The agaves in the left foreground are a small part of the lush garden of native plants with which Richard surrounded the house. The car, presumably Richard's, is an early 1950s Daimler-Benz DB-18 Special Sports. Richard Felger archive.

Richard's mother, Sadie, and her second husband, Martin Gang, posing for a formal portrait, possibly at the time of their 1974 wedding. Richard Felger archive.

Sadie and Martin Gang at a formal event, photographed with former U.S. president Gerald Ford (*left*) and actor-comedian Bob Hope (*right*), probably 1980s. Courtesy of Silke Schneider.

lived at 112 South Highland, Los Angeles [a bit east of Beverly Hills]. In 1936, my parents built the house I grew up in, 117 South Las Palmas—two blocks east of the Highland house. . . . The Las Palmas house was magnificent [you can find it online]. My mother did a lot of entertaining and was very social. At Hillcrest Country Club where [my parents] played golf . . . , the Sunday dinners were amazing."[8] Louie Felger and Martin Gang were quite social, too, hosting parties for Hollywood stars and socialites, presumably some of them their clients. Richard's family rented a summer home on the beach in Malibu, and while he worked at the Natural History Museum of Los Angeles County, he lived an hour's drive west of the museum near the beach.

Richard's paternal grandfather was Solomon "Sam" Felger, born May 12, 1870, in Odessa, Russia (Ukraine). He attended college for two years. He died February 29, 1944, in Los Angeles.

His paternal grandmother, Eva Bernstein, was born July 15, 1874, in Odessa, and died August 13, 1946, in Los Angeles. She graduated from the Imperial Conservatory of Music of Moscow, Russia, and attended the College of Music of Cincinnati, Ohio.[9] She later gave piano lessons in Bisbee and Yuma, as well as occasional recitals wherever she lived. Sam and Eva married in 1891 and migrated from Odessa to the United States in 1892. Family history suggests that the couple left Odessa because Solomon's parents did not approve of the marriage.

Richard wanted us to know, "Sam and Eva were most certainly not religious, [but] my parents were members of Wilshire Boulevard Temple. Rabbi Magnin was a friend, more of a Hillcrest [Country Club] acquaintance. Sometimes my mom would make us all go to one of the high holiday doings—more of a fashion show than anything else, but my dad and I were bored as hell. I do not remember Sam or Eva ever going to temple."[10]

As a child, Richard spent many hours with grandfather Sam and undoubtedly absorbed Sam's stories of wild events on the U.S.-Mexico border,

8. Richard Felger email to Karen Glikman, May 7, 2010.

9. *Bisbee (Ariz.) Daily Review*, February 22, 1903.

10. Felger email to Glikman, May 7, 2010.

Richard's paternal grandparents, Solomon and Eva Felger (*center*), with brothers Aron Glikman (*left*) and Samuel Glikman (*right*), 1939. The brothers were sons of Solomon's niece Chana Felger Glikman, making them Richard's second cousins. Solomon sponsored the brothers' immigration to the United States and funded Aron's first business venture. Courtesy of Karen Glikman.

clandestine trips to Europe, and days on the Arizona frontier running a dry goods store before statehood. Sam was a well-respected businessman who had risen from abject poverty as an immigrant to comfortable wealth as an entrepreneur. In a world of wars and persecution, bumps and bruises, he had seen it all and done it all. And prospered. We imagine that Sam was dapper, content, and a voice of wisdom in the family. Richard called him "gentle sweet Sam" and sounded very fond of him.

According to Richard, "Sam and Eva left Russia apparently as teenage lovers who married, went to New York, and headed west—I think for San Francisco. [They] stopped in Cincinnati, where my father was born, and then continued on west."[11] Sam worked along the way, making cloaks, running a lunch stand, selling clothes, and opening several dry goods stores. In the

11. Felger email to Glikman, May 7, 2010.

Richard's beloved grandfather Solomon "Sam" Felger, February 1, 1932. At some point Solomon adopted the name Samuel to memorialize his brother Samuel, who had been murdered in Russia. Courtesy of Karen Glikman.

mid-1890s the Felger family lived in Butte, Montana, where they undoubtedly met the Pincus family, but by 1899 documents show Sam, Eva, and Louie living in Arizona at Chloride (near Kingman), Williams in 1900, Bisbee by 1903, Yuma in 1905–8, and Douglas in 1910. Between 1910 and 1912, Solomon opened a general store on the international border at Naco, Arizona (just south of Bisbee). Naco is a sister or twin "city"—village really—to Naco, Sonora.

One of his ads read, "The Law Prohibits Nakedness. We have the best assorted stock in ready-made clothing, shoes, hats, etc."[12] Richard believed that Sam and Eva "stopped in Naco because . . . it was a beautiful area and the mines and commerce were booming,"[13] a perfect place for a store.

"Their mercantile establishment was right on the border; there was no fence between Sonora, Mexico, and Arizona in those days. My grandfather was well regarded as a gentleman and respectable when they lived in Naco,

12. *Mohave County Miner*, November 11, 1899.

13. Felger email to Glikman, May 7, 2010.

but I do know . . . that gentle sweet Sam also sold arms to Mexican factions during the Mexican Revolution, so he did have another side."[14]

"During the Arizona years, Sam and Eva went to Europe on occasion—I have no idea how often. One year they had their steamship trunk all packed . . . , and Pancho Villa drove his wagons up to the side of my grandparent's place and took EVERYTHING they owned, including the steamship trunk that was conveniently already packed. I heard that my grandmother was especially upset because they even took her underwear—everything in the house and store was taken. They did not go to Europe that year, as they were wiped out. Did they report this to the authorities? I doubt it."[15]

In 1915, Pancho Villa did briefly take control of Naco, USA, but apparently Richard's grandfather rebounded financially, for in 1918 he led Naco's honor roll for buying government Liberty Bonds worth $6,000 to help the war effort. The second highest on the list was for $1,000.[16] Sam and Eva's only child, Richard's father Captain Louie Felger, was serving in France as an army medical doctor.

Richard believed that Sam and Eva's "trips to Europe might have been to dispose of French money. They exchanged Mexican money with French Jews in Nacozari [Mexico], the rich mining town a day's stage ride south of Naco, Arizona."[17] France had occupied Mexico from 1861 to 1867, but French money may have been shunned thereafter, making it worthless in Mexico. "Such were the times," Richard wrote, "and few people trusted banks and government."[18] Sam must have been comfortable carrying large sums with him to exchange for the best rates.

"And I am pretty sure my grandfather illicitly got money to Switzerland to buy people out of the [Nazi] death camps—although that was very much against wartime law."[19] Sam and Eva practiced *mishpoche*, a Yiddish concept meaning "extended family," but also covering the practice of bringing Jewish

14. Felger email to Glikman, June 6, 2011.

15. Felger email to Glikman, May 7, 2010.

16. *Bisbee (Ariz.) Daily Review*, April 28, 1918.

17. Felger email to Glikman, May 7, 2010. One of Sam Felger's business contacts in Nacozari was named Schwab or Schwoeb.

18. Felger email to Glikman, June 6, 2011.

19. Felger email to Glikman, June 6, 2011.

family members or friends to the United States.[20] They sponsored Chana (Felger) Glikman's three sons starting in 1930, followed by Chana and her daughter in 1940.[21]

After they left dusty Arizona for good, perhaps in 1922, Sam and Eva moved to Los Angeles and later "lived at The Ravenswood Apartments, on Rossmore Avenue in Hancock Park, close to Las Palmas and the Larchmont shopping area. Mae West owned The Ravenswood and lived there from 1932 on.[22] There were extensive gardens on the north and south side."[23] And it was not far from their son's homes on Highland and then on Las Palmas.

"Sam Felger was retired the whole time I knew him. . . . His close friend, and I think the friendship went way back, was named Karp. They played cards together, usually sitting in The Ravenswood garden. They called each other 'Felger' and 'Karp'—never first names. I thought it was funny that the old man was named after a fish."[24] At Sam and Karp's oldest ages and a decade beyond, Richard was not playing cards—he was giving talks, writing books, and hiking mountains in search of plants.

Had their lives overlapped longer, we can imagine that in hushed voices devoid of bravado, Sam, the gentle cloak maker, and Richard, the peaceful plant presser, might have traded discreet stories comparing Villa's raid on the Felger store with Richard's deeply traumatic experiences evading armed drug smugglers and bandits in Sinaloa and Sonora. Or, in retrospect, they might have confessed their fears or sighed with relief over long-ago missions: Sam's covert currency missions to and from Europe, and selling guns to armed rebels along the border; or Richard's botanizing near cartel-guarded poppy fields, or marching in rebellious protests, or tumbling down a steep slope toward a cliff and grabbing a cholla cactus just before he would have pitched over the edge

20. Abraham S. Chanin, *Cholent and Choriza: Great Adventures of Pioneer Jews on the Arizona Frontier, Sometimes Kosher, Sometimes Not, But Always Fascinating* (Tucson, Ariz.: Midbar Press, 1995).

21. Glikman Family History, courtesy Karen Glikman.

22. Mary Jane "Mae" West, who made a fortune as a famous actress, singer, comedienne, and sex symbol, once said, "You only live once, but if you do it right, once is enough." The chesty Mae West life vest was named in her honor.

23. Felger email to Glikman, May 7, 2010.

24. Felger email to Glikman, May 7, 2010.

to his death below. During their clandestine operations and harrowing escapes, both men had faced imminent death, robbery, or arrest. In one terrifying incident that Richard did not write about, he was robbed at pistol point by two thieves on the Arroyo Guajaray in the Sierra Madre of Mexico. Too, both he and Sam had careers where they gambled on the next store or next grant, so going broke always loomed on their horizons. Sam, playing cards with his old friend Karp, and Richard, quietly sorting sheets of dried plants, would seem to be an unlikely pair of adventurers, but they were.

One still unexplained story was the education of Richard's father, Louie. Moving from boom town to boom town with the family store would have made schooling difficult, so by 1907, when Louie was about fifteen, he was taken to Los Angeles, where he finished high school, and then graduated from college and medical school at USC.[25] Since his parents had remained in Arizona, he may have stayed with Abe and Minnie Karp, close family friends.[26] By 1924, son Louie was on his own, having served with the army in Europe and visited Japan and China.

"After my grandfather died," Richard wrote, "Eva only lived another year or so—she was heartbroken and lonely. At the time I often stayed overnight with her at Ravenswood. She played piano all day almost without stopping. I was mesmerized just listening to her, maybe the only person permitted to listen. As I paged through my reptile, marine biology, and botanical books, I would read . . . and listen to her."[27]

Richard never met his maternal grandparents. They were Adolph Pincus and Hattie Lewkowitz Pincus. Adolph was born November 27, 1859, in Kempren, Posen, Germany (Poland), and he died in 1929 at Butte, Montana. Hattie was

25. "Louis Felger, son of Mr. and Mrs. Sam Felger, will not return from the coast, but will enter the high school in Los Angeles," *Arizona Sentinel*, July 31, 1907. One fragment of family history states that Louie attended school in Urbana, Illinois, in 1909 while his parents lived in Douglas. Possibly the school was University of Illinois Urbana-Champaign, which opened in 1868. Louie was age seventeen in 1909.

26. Karen Glikman email to Bill Broyles, August 8, 2024.

27. Felger email to Glikman, June 1, 2010.

born in December 1869, in Berlin, Germany, and she died June 29, 1932, in Butte. Hattie and Adolph emigrated separately from Germany to the United States in 1881, and they married in Butte in 1889. They had two daughters, Sadie Dorothy and Frieda. Frieda first married Jesse Clarence Levy (1896–1963) and had two children, Joan and Nancy. Her second marriage was to Barnard C. "Barney" Bristow, and they lived in San Mateo, California.

Adolph's brother, Isadore (1855–1912), had immigrated to the United States in 1871 or 1875 and had settled in Butte by 1878. Adolph moved to Butte in 1886, and then on to Clark County, Washington in 1887 where he worked as a laborer.[28] In 1889, the year that Adolph and Hattie married, Adolph bought mining claims from his brother.[29] By 1896 Adolph owned and ran I. M. Jacobs and Co. on what was called the "Pincus Block" at 26 South Main in Butte.[30]

Isadore worked as a travel agent and had a cigar business in Butte, but the brothers feuded over accusations of gambling and fraud, followed by a lawsuit, and ended with Adolph publicly denouncing his brother's gambling. Isadore and his wife, Bertha, sued Adolph, accusing him of "swiping" Isadore's cigar business.[31] In court Adolph testified that Isadore "was no good, that he would have no more to do with him." The court found for Adolph, and the rift did not heal.

By 1898 Adolph described himself as a "retired capitalist," and in 1900, when the U.S. Census listed him as a merchant, the family employed two servants.[32] In 1910 Adolph was listed as a jeweler, and in 1920 his occupation was real estate agent.[33] He died in 1929, and his will stipulated specific bequests of more than $82,000 and indicated an estate worth about $500,000, including realty holdings in Montana and California.[34]

Richard Felger's parents and grandparents lived the American dream. They fled religious and ethnic persecution in eastern Europe and sailed to

28. *Butte (Mont.) Semi-Weekly Miner*, February 17, 1886.

29. *Anaconda Standard* (Montana), December 13, 1889.

30. The 1896 Butte, Montana, city directory

31. *Anaconda Standard* (Montana), February 29, 1896.

32. *Anaconda Standard* (Montana), May 24, 1898.

33. 1910 and 1920 U.S. Census records.

34. "Will of Pincus Filed in Court," probably in the *Montana Standard* (Butte), 1929. Date and page missing from news clipping.

America, where they began their rags to riches story. Crossing the continent, they lived in rough and tumble mining towns, found work, established stores and offices, and finally settled in California, the Golden State, where first- or second-generation Americans could become prosperous businessmen, doctors, or entertainers. As for Richard, he followed his passion for wild plants.

## FAMILY AND SCHOOL

Richard Stephen Felger was born in Los Angeles, California, on August 19, 1934. He was the son of Louie Felger, a physician, and Sadie Dorothy (Pincus) Felger, who oversaw daily or live-in household help that nannied, cleaned, cooked, gardened, and chauffeured.[35] His brother, Dan A. "Danny" Felger (1930–2010), was a successful California real estate developer and investor married first to Nellie Donivon Hailey and then to Carolyn Mae Spicer. Dan enjoyed tennis and ocean sport fishing.[36] Dan is the paternal grandfather and Richard the great-uncle of American actress Mikey Madison, who won the 2025 Academy Award for Best Actress for her role in the movie *Anora*.

Richard's family lived in Los Angeles, where he attended elementary, junior high, and high school, presumably from 1940 to 1952.[37] He enrolled at the University of Arizona in the fall of 1952, received a BA in 1956, an MA in the spring of 1959, and his PhD in 1966, all at Arizona.[38] His master's thesis was "Clinal Variation in the Surface-Volume Relationships of *Lophocereus schottii* in Sonora, Mexico." His ambitious dissertation was "Ecology of the Gulf Coast and Islands of Sonora, Mexico," which later served as a foundation for his book *Plant Life of a Desert Archipelago*, with Benjamin T. Wilder and Humberto

35. Family maids included Emma Trixal (or Trixel), Maybelle Burdett, and Eunice Rainey (courtesy Sue Rutman).

36. "Dan Felger Obituary," accessed June 14, 2024, www.legacy.com/obituaries/name/dan-felger; *Los Angeles Times* obituary, November 22 and 23, 2010; *Chum Line* newsletter, Los Angeles Rod and Reel Club, December 2010.

37. He attended John Burroughs Middle School in Los Angeles.

38. Various sources: University of Arizona Yearbook, 1953, 1954, 1955, 1956, available at University of Arizona Main Library website; "Dean's Honor List, University of Arizona," *Tucson Daily Citizen*, March 8, 1954—Felger was a sophomore in 1954.

Richard at the wheel of an early 1950s Daimler-Benz DB-18 Special Sports, with the steering wheel on the right, November 1955. Though presumably his car, it would have been impractical for fieldwork in Mexico. The car he drove on his first field trip in 1952 or 1953 was a 1951 Ford, to be followed soon thereafter by an early 1950s Buick. Apparently, for many years he kept both a town car and a separate vehicle for fieldwork. Courtesy of Silke Schneider.

Romero Morales. Felger's dissertation was submitted to the Zoology Department, and his dissertation committee consisted of his adviser Charles H. Lowe, Charles T. Mason Jr., Howard K. Gloyd, and two other professors. Lowe was also his master's adviser.

As an undergraduate at the University of Arizona, Dick Felger was a member of the Zeta Beta Tau fraternity from 1952 to 1956. At that time it was an all-Jewish group with the credo "Intellectual Awareness, Social Responsibility, Integrity and Brotherly Love." He does not appear in other campus clubs or activities in the school yearbooks as an undergraduate, though the 1954 edition

Richard's Mercedes-Benz 300SL Gullwing, pictured here in Colorado in 1967. Depending on rear-axle gearing, they were capable of top speeds up to 163 mph. They were only made from 1954 to 1957. Richard Felger archive.

noted that he earned a 1.0 (A) grade average. Sometime in the 1970s, he began calling himself Richard.

As a graduate student, Richard showed a bit of a wild streak that belied his reputation as a quiet scholar when he raced his Mercedes-Benz 300SL Gullwing sportscar against a countess who also owned a 300SL. Richard won, pocketing the $50 bet. A fantasy, you wonder? The lady was Margaret "Daisy" Howard (1879–1968), Countess of Suffolk, a wealthy Chicago socialite who had married British nobility, and after her husband died she moved to Tucson, bought land northwest of town, and lived famously. One of her parcels became the Suffolk Hills neighborhood, and her 3,500-acre ranch with its distinctive headquarters near Oracle ended up as Biosphere 2. Her favorite car was a Bentley or Rolls Royce, at times driven by her chauffeur, a former Royal Air Force pilot.39

39. See "Lady Suffolk Makes a Life in Tucson and Oracle," Oro Valley Historical Society, accessed December 13, 2025, https://ovhistory.org/in-search-of-the-countess-lady-suffolk/;

## CAREER

Richard's parents probably wanted him to become a medical doctor, like his father, but even before he started grade school, he became enchanted by creatures of the shore and sea—fishes, mollusks, seabirds, and kelp. By age eight, he had declared himself to be a botanist. He was an early reader and hounded his indulgent father to buy serious science books for him. The world of the Sonoran Desert opened up to him on a high school field trip with his beloved teacher Nancy Thomas Neely and her husband, Peter. Trips to Organ Pipe Cactus National Monument in Arizona and then Álamos, Sonora, showed him the "good stuff," as he called it, and the rest became history. The University of Arizona in Tucson offered an ideal base for his lifetime of plant studies and his curiosity about the desert and its Native peoples.

Following grad school, his career began in the fall of 1966 as an assistant professor in the Department of Biology at the University of Colorado, but he left teaching in 1967 after just one year. The suspicion is that he wanted to be outdoors following the cycles of blooming desert plants or boating with Seris from island to island instead of meeting class schedules, grading exams, or attending faculty meetings. Throughout his life, however, he became an increasingly patient and encouraging teacher who informally taught, guided, and inspired young botanists, Indigenous naturalists, hobbyists, late-blooming scholars, and even peers. He enjoyed leading groups on show-and-tell plant walks, often sharing his hand lens to show them something special about a particular plant. His sense of wonder was contagious, and even with children he could explain plant characteristics and answer questions in understandable terms.

From 1967 to 1969 he served as senior curator of botany at the Natural History Museum of Los Angeles County in Long Beach, California, also with a title as adjunct associate professor, Department of Biology, University of Southern California, 1968–69. During this time he joined the Population Crisis Committee and, along with Paul Ehrlich, author of *The Population Bomb*, publicly advocated zero population growth.[40] And while teaching in Colorado

Wikipedia, "Margaret Howard, Countess of Suffolk," last edited on December 7, 2025, https://en.wikipedia.org/wiki/Margaret_Howard,_Countess_of_Suffolk.

40. *Long Beach (Calif.) Press-Telegram*, October 24, 1968.

and later working at the museum in Los Angeles, he participated in marches against the war in Vietnam.

He did not stay away from the desert for long, and when he returned to Tucson, he settled into a series of positions or titles, though we're unclear what his duties were or if the jobs paid anything: 1969–70, visiting scientist, Department of Geochronology, University of Arizona; 1970–71, consultant to the Arizona-Sonora Desert Museum; 1969–75, research associate, Natural History Museum of Los Angeles County. Also in 1973, he worked at the Environmental Research Lab, University of Arizona, perhaps as a consultant.[41] Commonly, adjunct instructors are paid to teach a class or two per semester, but without job benefits or the teaching counting toward tenure.

He may have been cobbling together part-time jobs or receiving family help, but his personal and social life took unpredictable turns. For example, about 1973 he was in Bolinas, California, a reclusive artist colony northwest of San Francisco, looking for someone to type a manuscript, when he met Mahina Drees. They became a couple for the next four years and even filed joint income tax forms. They lived at Tucson in one of several rustic low-rent rock houses at the mouth of Esperero Canyon, just east of the Flying V Guest Ranch (now Loews Ventana Canyon Resort) in the Catalina Foothills north of Tucson.[42]

To make ends meet financially, Drees took temporary jobs, while Felger applied for grants, may have had some part-time work, or may have dipped into savings from selling a previous home. Drees was an actress, folk singer, song writer, and later co-founder of Native Seeds/SEARCH, based in Tucson. In her words, "Richard made plants fun."[43]

Also living at Esperero Canyon off and on were Andy Weil, Peter Warshall, Gary Nabhan, Ron Pulliam, and Ed Abbey, as they were known before they became famous.[44] Later Andrew Weil became a well-known physician and advocate for alternative medicine, Peter Warshall an ecologist and biogladiator,

41. *Long Beach (Calif.) Independent*, August 3, 1973.

42. The property was homesteaded by Randolph "Pat" Jenks in the 1930s. See Jenks's *Leaving the Golden Age of the 1920s for Adventures in the West* (Tucson: Arizona Lithographers, 2003), 61–68.

43. Mahina Drees interview with Ben Wilder and Bill Broyles, Tucson, June 14, 2024.

44. Felger had met Warshall and his partner, Joanne Kyger, a poet, in Bolinas.

Gary Paul Nabhan an award-winning botanist and writer, H. Ronald Pulliam an influential conservation ecologist and science adviser to the secretary of interior, and Edward Abbey a writer and voice for wild places. Among Richard and Mahina's circle of friends were Diana and Drum Hadley (she was an ethnohistorian, and he a rancher and poet); heiress Peggy Mellon Hitchcock (Richard taught her son the multiplication tables); Beat poet Allen Ginsberg, who was writing letters to Congress about global warming as far back as 1969; and Ann and Bill Woodin, she a writer and he director of the Arizona-Sonora Desert Museum, where Richard later worked.[45] Today this group reads like an all-star team of naturalists, environmental leaders, writers, and advocates, with Richard standing as tall as any of them.[46]

After Richard and Mary Beck Moser's book *People of the Desert and Sea: Ethnobotany of the Seri Indians* was published, he sent copies to amicable friends, including Ginsberg, who replied,

> Dear Richard: received yr. monumental volume—what a knockout, lot of work, wisdom, observation, art, paths, ways, knowledge, myths, flowers and cacti, language, animals, shelters, tides, roots, huts, history, fetishes, medicines, & cuisine—dazzling knowledge saved for some human use when the tornado of fossil fuel and steel passes over if ever—wish I had more time to absorb all that fact and poetry, but just leafing through reading spots, captions and paragraphs is a thrill and awesome. Congratulations & many thanks for thinking of me and sending a copy for me to look at in leisure. . . . Admiration! Allen Ginsberg.[47]

45. Naomi Oreskes, "Gaslighting by the Gas Industry," *Scientific American*, July/August 2024, 96.

46. One could write a book about the lives of the Esperero Canyon crew and their friends, and how they each followed their hearts to study and save the world. Diana Hadley was in the Woolworth family; Drum Hadley was a Busch, as in Anheuser Busch; and living-museum director Bill Woodin was the son of an industrialist who became secretary of the treasury under Franklin D. Roosevelt.

47. Allen Ginsberg postcard to Richard Felger, August 10, 1992. In a concurrence of spirit, the postcard was produced by "A Peace of Mail. Man and Nature on Paper Gallery."

Most of Felger's fourteen academic papers published during 1967–1974 stemmed from his master's thesis on *senita* cactus and research among the Seris, with topics including the nutrition of eelgrass and Seri uses of mesquite and of agave. His acclaimed 1976 paper on sea turtle dormancy in the Gulf of California was especially rewarding for the boy who once intended to be a marine biologist. These articles appeared in the globally known journals *Science* and *Ecology*, as well as in the southwestern anthropology and history journal *Kiva*, but none was lucrative or clearly tied to a job.

Nonetheless, he was building relationships with colleagues, institutions, and undoubtedly funders. From 1974 to 1977, he was a research associate at the Arizona-Sonora Desert Museum in Tucson, and from 1977 to 1979, an adjunct associate professor, Department of Ecology and Evolutionary Biology, University of Arizona, which may have been a paying position. Then in 1978 he landed a new and significant position as ecologist, senior research scientist, and head of the research department at the Arizona-Sonora Desert Museum. Basically, he established the museum's still-thriving research department. One person who also worked at the museum back then said that Richard had a staff of two, a secretary and a computer tech, and he kept both of them very busy—the secretary, Lynn Ratener, even co-wrote a paper with him. He survived policy skirmishes at the museum but left in 1982, at least in part because he was spending much of his time in Mexico researching sea turtles and studying the Seris.

Apparently the museum experience led him to quit working for others altogether and to become his own boss, self-employed and full time. He foresaw his future built on grants and contracts that would allow him to do his own research, write papers, and contribute energy and words to his favorite causes of human population control, identifying dryland plants as potential crop species, protecting native plants, mapping plant distributions, and understanding how Native Americans used plants. His first major grant came from the Ed Janss Foundation in Los Angeles for his Seri ethnobotany work; the National Science Foundation (NSF) and Roy Chapman Andrews Fund at the Desert Museum also provided significant support.

Another NSF grant in the 1970s allowed Felger to study historic photo archives and artifact collections of the Seris for his forthcoming ethnobotany book. The museums and libraries he visited included the Smithsonian Institution, Heye Foundation, Museo de Antropología in Mexico City, Peabody Museum, National Museum of Ethnology in Osaka, Japan, and seven other institutions.

Richard with the species *Verbesina felgeri* in Cañón del Nacapule, October 2015. Photo by Jim Malusa.

Working on grants requires rigorous self-discipline, and living on grant money requires courageous leaps of faith that paychecks will arrive before the rent comes due. As Richard explained to a friend in an email, "we're hanging on a thread waiting for renewal check for the big grant—flat out of money—the grant has been approved—and the check usually gets here soon—if it doesn't—may we squat on your estate?"[48]

At times, he worked on his own projects under general unrestricted support from a patron. Richard had several reliable benefactors, including Henry

48. Richard Felger email to Bill Broyles, November 2, 1998.

Browne "H. B." Wallace and Linda Wallace Gray of the Wallace Research Foundation and Wallace Genetic Foundation, as well as Jane Harrison Ivancovich, a Tucson patron of culture and history. Peggy Hitchcock and Gil Gillenwater also strongly supported Richard's research. One friend from the 1970s recalls Richard saying that comedian and family friend Bob Hope regularly underwrote some of his work, as did his parents.

His books came later, but sales of his books paid virtually nothing, and some journals where he published even required researchers to pay page charges for publication. But as his wife Silke Schneider told us, "Richard was very successful at getting grants, and those supported his work. He had some major supporters over the years." She did not know of any Felger family inheritances, though two other friends from an earlier time in Richard's life believed that his mother had supplemented his income, and Silke did say that his mother contributed to buying his Isuzu Trooper.

A favorite of his activities was developing floras (plant lists) for various regions, including fossil plants found in packrat middens by paleobotanist Tom Van Devender and others, dating back thousands of years. He thoroughly enjoyed identifying species, and often he crashed past deadlines while looking for one more specimen in his study area. Friends and co-authors had to tell him, "The train is leaving the station" or "Stop!" so that he would wrap up the manuscript and send it for publication. Like a diligent collector looking for the very last coin in a series, Richard craved a full set. His tens of thousands of plant specimens can be found in an array of herbaria around the world. Richard himself described several new species, and three new species were named for him by other botanists: *Agave felgeri*, *Isocoma felgeri*, and *Verbesina felgeri*.

At the University of Arizona Herbarium, where he hunched over a table to examine details of stamens, leaves, and nutlets, he tolerated short questions from other researchers or students—but he declined chatter or gossip, and he deferred to those he most admired, like the Reeders, John and Charlotte, who seemed, he said, to have studied grasses since the dawn of time. He expressed awe for scientific illustrators like Lucretia Brezeale Hamilton, Matthew Johnson, Jeanne Janish, and Cathy Moser Marlett, who could accurately—and artfully—depict the essence of leaves, flowers, and seeds in pen and ink.

Richard had several goals in the herbarium. First, he insisted on confirming past species records for an area to his own satisfaction before publishing its presence in his flora—he checked everyone's homework, even his own. He loved chasing down doubtful, enigmatic, or potential records, such as when a specimen was mislabeled or misidentified, or a species had been split into subspecies, or subspecies combined into one. Second, he almost always constructed his own unique identification keys for an area, a difficult meta-logic process that eased the time and effort for others to call out a plant's name. Third, he sometimes found hybrids and crosses of species, as if seeing evolution in action. Fourth, he wanted to see how species' ranges were expanding or contracting, or whether invasive species were on the move or crowding out natives. And fifth, he was making sure that his own donations of specimens, *his* gifts to posterity, were accurate. If Richard dreaded one thing in his life, it was for him to be wrong.

It is important to remember that field botany is not the solitary enterprise it might seem, for it requires constant collaboration and updating of species names, coordination of information, financial underwriting, and cooperation in placing specimens at herbaria and in publishing results. To these ends, Richard seems to have knocked on every door seeking support for his work. It's a demanding career based on trust, expertise, and hard-earned results.

Richard found advantages in linking up with various academic institutions as a "research associate." That label opened doors on campuses and buoyed his credibility on grant applications, and sometimes an institution or non-profit group managed his grant money for him to their mutual benefit—Richard detested the chore of bookkeeping.

From 1980 to 2002, Richard's links included adjunct senior research scientist in the Department of Soil, Water, and Environmental Science (formerly the Environmental Research Laboratory) at the University of Arizona. He was a research associate for the University of Arizona's Office of Arid Lands Studies from 1981 to 1985, and he wrangled recognition as an *investigador invitado* at the Instituto de Ecología in Mexico City from 1984 to 1986. Felger continued to receive grants, and he retained his academic links, even adding a few, including *asesor científico* of the Estación Regional Hermosillo del Centro de Ecología at the Universidad Nacional Autónoma de México in Hermosillo, Sonora (1990–2000); and *revisor permanente*, Consejo Editorial del Ecológica at the Centro de Ecológico de Sonora in Hermosillo (1992–2000). Both of these Hermosillo connections likely were facilitated by his botanist friends Alberto Búrquez and Angelina Martínez-Yrízar.

Richard and a dwarf palm tree in the garden of the Taj Mahal, south of New Delhi, during a trip to the Thar Desert of northwestern India with James Aronson, March 1982. Photo by James Aronson.

From about 1981 to 1987, Richard worked as an international plant hunter, a detective with a hand lens. Under the Environmental Research Laboratory's auspice, he teamed with James Aronson to search for xerophytes and halophytes of known or potential economic interest. Aronson worked for the Hugo Boyko Applied Research Center in Beer Sheva, Israel. Together they visited Israel's Negev and Egypt's Sinai Deserts, the hyperarid Atacama Desert of northern coastal Chile and the rain shadow desert of El Monte in Argentina, the Guajira Peninsula of Venezuela, Baja California, and the Thar Desert of Rajasthan in northwestern India.

Then, in 1988 he established his own non-profit corporation and became founder and executive director of Drylands Institute, Tucson. Its mission read "Drylands Institute is a non-profit 501(c)(3) research/education organization based in Tucson, Arizona. We have a commitment to advancing knowledge of southwestern North America's flora, fauna, ethnobiology and the region's human inhabitants. Our research efforts focus on the Sonoran Desert and select arid and semi-arid regions of the world. We provide information that supports the conservation of natural and cultural resources such as the preservation of biological diversity including rare breeds of domestic animals and improvement of health and economic welfare of people."[49] Richard positioned Drylands to identify, inventory, and interpret biological diversity as building blocks of conservation efforts, and then to publish and present those findings for use by scientists, agencies, and advocacy groups.[50] His heartfelt mission was nothing less than to save the world.

Several capable biologist-researchers handled Drylands' office chores over the years, including Dale Turner, Linda Leigh, Jared McKinley, and Michael Wilson. Drylands was Richard's business vehicle until 2007, when he relinquished its leadership to Wilson and moved to Silver City, New Mexico, where Adrienne Booth helped him find and secure grants, as well as create his website.

He became a research associate at the San Diego Natural History Museum in 2007 when Exequiel Ezcurra became provost there, and Richard became a visiting scholar at Western New Mexico University, Silver City, when he moved there in 2007. These relationships also linked by internet an international brain-trust network of other researchers working on a variety of projects, and in the course of a day, Richard might exchange flurries of questions and answers with a score or more botanical or other scientific experts around the world.

And he renewed his research association with Sky Island Alliance in Tucson from 2011 to 2015. He had been a co-founder of this cross-border environmental advocacy, research, and outreach group in 1991—its first meeting was at his house, and he was one of three who signed its non-profit incorporation application in 1995, though he left its management to others. Also, Richard was a mentor and cheerleader in 2012 for formation of a U.S.-Mexico coalition of researchers and leaders organized as Next Generation Sonoran Desert Researchers (N-Gen). By 2024 N-Gen had more than one thousand members and a very active agenda promoting research and collaboration.

49. Drylands Institute, Arizona incorporation charter. The rare or heritage breeds section was added after he met Silke Schneider; her dream project was a Heritage Breeds Farm Park.

50. "Information Resources Supporting Conservation of the Sonoran Desert Region," grant proposal to the G. A. Binney Conservation Foundation, about 2005.

After a day in the field, Richard and Ben Wilder (*right*) identify and prepare plant specimens in Steve and Cathy Marlett's rental house in Bahía de Kino, Sonora, September 2007. Photo by Cathy M. Marlett.

For someone who abandoned the classroom after one year of classroom teaching, Richard distinctly valued his own education and generously mentored a couple dozen younger men and women over the years. He served on their PhD committees, visited their field sites, traded continual emails and phone calls, reviewed drafts of their papers and grant applications, and helped them identify plants and compare herbarium specimens. Moreover, his "office" was always open, and he provided boundless encouragement. He wanted them to succeed, and he yearned to know the answers to their questions. Typical was his remark to Benjamin T. Wilder, then an undergraduate student interested in researching Gulf of California islands: "Why don't we go there, fill in some gaps, do the flora, and write a book." In essence, he helped grow a generation of reliable peers who in turn helped him and each other.

In the 1970s, Richard had penned several articles for Defenders of Wildlife, including a report on the 1972 United Nations Conference on the Human Environment he attended in Stockholm, which was the first world conference to make the environment a major issue. That conference adopted the Stockholm Declaration and Action Plan for the Human Environment, as well as more than two dozen resolutions. In 1978 he attended the Fourteenth General Assembly of the International Union for Conservation of Nature (IUCN) in Ashkhabad, USSR.

Closer to home, Richard wrote and spoke in support of Stewart Udall's proposal for a Sonoran Desert International Park in 1966 and again when the idea resurfaced in the late 1990s. Richard actively supported creation of the Sonoran Desert National Monument that was established in 2001, and in 2002 he suggested an Escalera Ecológica concept for the Gulf of California to pair with the Mexican government's Escalera Náutica plan, his intent being to counterbalance the government's increased emphasis on port-to-port Gulf tourism by adding a "ladder" of ecological areas and environmental protections. He spoke in favor of a corridor of reserves throughout the Gulf of California and northwestern Mexico during a keynote presentation for the Fifth Baja California Botanical Symposium at the San Diego Museum of Natural History in July 2002. Since then, a number of Gulf islands and coastal lands have been designated as UNESCO biospheres, and Richard's voice, published papers, and books served as cornerstones for those decrees.

As he reached his seventies, he imagined a legacy conference and book that would revisit and reinforce his earliest themes and reintroduce them to an awakening worldwide audience. He proposed a four-hour panel and speaker session for the 2007 Society for Conservation Biology Convention, then being planned for the Nelson Mandela Metropolitan University at Port Elizabeth, South Africa. The session would document and explore the historical correlation of population increases with environmental stresses and wars. He sought to sound the alarms that unchecked human population growth erodes and threatens sustainable economies, biodiversity, and intercultural respect, endangering nature itself, including our own human nature.[51]

51. The tentative list of speakers included, with their affiliations at that time, Elizabeth Arnold (University of Arizona Department of Plant Sciences), Sylvia Earle (DOER Marine Subsea Robotics and Submersible Systems), Exequiel Ezcurra (San Diego Museum of Natural History), Thomas E. Lovejoy (The H. John Heinz III Center for Science),

In conjunction with the conference, he also proposed a book based on the model of Jean-Luc Cartron, Gerardo Ceballos, and Richard's book *Biodiversity, Ecosystems, and Conservation in Northern Mexico* (2005) and Richard and Bill Broyles's *Dry Borders: Great Natural Reserves of the Sonoran Desert* (2007). And he proposed a perennial website. But he found only lukewarm financial support for his suggested budget of $275,000 over two years. Although the project would have been a crowning summation of his multifaceted career, a full-thrust effort would have diverted him from his other worthy projects already underway. With regret, he laid the proposal aside.

Richard wrote, edited, or co-authored fourteen books, including this one in your hand. Remarkably, his first book, *Dark Horses and Little Turtles* in 1974, was poetry, not science-styled formal prose, but it announced his larger lifelong themes of Gaia, ecology, evolution, and feeding humanity. It appeared two years after he attended the Stockholm conference. Across the arc of his life he wrote more than one hundred peer-reviewed papers, as well as numerous popular essays about plants, resource sustainability, and Native peoples. The number of his grants and quality of that work can be gauged by his prodigious publication list and the academic standings of his co-authors.

He neither held nor attended his own retirement party. Why? Because he never retired. He wrote until his very last days, and he would be relieved to know that two of his book manuscripts have been finished and published posthumously, plus his autobiographical stories are now the core of this book, *The Botanical Odyssey of Richard S. Felger*. His archive, field notes, and publications will aid researchers for generations to come.

## HOME

Richard married three times that we know of: first to Patricia Lou Seff (July 21, 1956, in Denver), second to Ruth Courtland (September 8, 1977, in Tucson), and then Silke Schneider (2009 in Silver City).[52] The first two marriages ended

Wallace J. Nichols (The Ocean Conservancy), Gary Paul Nabhan (Center for Sustainable Environments), Peter H. Raven (Missouri Botanical Garden), and Richard himself. Other experts from around the world would have been invited.

52. Patricia Lou Seff later married Barry W. Messer in 1965 (Ancestry.com).

in divorce. He had no children. Along the way, he did cherish a succession of live-in girlfriends, remarkable women each as intelligent and vivacious as the others. These partnerships ended amiably; as one told us, "Richard worked all the time and traveled a lot, so finally I went off to Oregon with a mutual friend. But Richard and I remained friends."

He lived in Tucson, Arizona, during his college years, 1952 to 1966, and again from 1969 to 2007. In 1994 he met Silke Schneider, who quickly became his soulmate. They connected at Biosphere 2 when Richard gave a poetry reading. Silke, known as "Safari" in the Biosphere 2 community, was intrigued.

One Valentine's Day, Richard gave Silke a cactus instead of roses, but she understood it came from his heart and appreciated it all the more. Together they'd sometimes attend movies, concerts, or presentations. But when they went for walks, they held hands and talked about everything under the sun. Silke was as serious about animals as Richard was about plants. In the circus, she had signed up to be an elephant rider and then was assigned a role as a horseback cowgirl in an Italian traveling show called The American Circus. Over several years she worked as a jillaroo (cowgirl), rounding up cattle in Western Australia (a "muster," they call it Down Under). In one notable publicity stunt, Caravan of Dreams or Bust, she camped cowboy-style and rode horseback 1,100 miles on the historic Santa Fe Trail from Santa Fe to Dodge City, Kansas, and then south on the old Chisholm Trail from Dodge City to downtown Fort Worth, Texas.

When she met Richard at Biosphere 2, she had co-designed the domestic animals part of the project and was the animal systems manager. After she left Biosphere 2, she studied animal science at the University of Arizona. Too, she had served two terms on the board of the Livestock Conservancy, a national group devoted to conserving rare breeds, and she was on the board of Rare Breeds International. Today she raises Spanish Barb horses, emus, and turkeys. You can easily see why Richard was as intrigued as she was. As she said, "We were both consumed by living our work, and we shared common values."[53]

Neither Richard nor Silke was fully understood by their families, but they found validation in each other's causes, acceptance, and love. Late in his career Richard attended a conference in Baja California Sur and was presented

53. Silke Schneider interview with Ben Wilder and Bill Broyles, Silver City, New Mexico, December 30, 2023.

Richard and Silke in the late afternoon sun on a beach near Guaymas, Sonora, January 1996. Photo by Gil Gillenwater.

Richard's house in the historic Sam Hughes neighborhood of Tucson, March 1992. Richard Felger archive.

a lifetime achievement award. On his return trip to Tucson, he swung past Los Angeles to visit his mother and show her his plaque. "Do you know what she said to me?" he asked one friend. "She said, 'Well, Richard, I'm glad you finally made something of yourself.'" Richard's wife Silke had faced the same misunderstanding with her own parents not fathoming their daughter running off to join the circus instead of going to college.

Richard was born and raised rich, but he didn't stay rich. Being rich was his parents' world. He sealed his fate at age eight with his decision to become a botanist instead of a physician to the stars, a big-time real estate developer, or a lawyer to the elite. He knew his way around money, but after he graduated

Richard and Silke on the deck of their Silver City house, late 2000s or early 2010s. Courtesy of Silke Schneider.

from college, his career pressing plants did not entail a lucrative or even steady income. He was committed to the life he chose and was content with it. As much as he loved his mother and she loved him, she never really understood him. More than Sunday lunches at the country club and hosting parties for Hollywood stars, he preferred sitting around a campfire with Seris fathoming the fate of sea turtles and pursuing obscure plants that could feed the world.

In Tucson Richard lived in several different homes that he later sold at profits. His Sam Hughes neighborhood house was only a short walk from the university herbarium, but Richard detested the neighbors' noisy leaf blowers, and after he and Silke met, they moved to a foothills ranchette in the unincorporated community of Catalina, north of Tucson. For a decade that seemed ideal, but the hectic drive through Tucson's city traffic going to and from the University of Arizona Herbarium proved too much for Richard, and a yearning for a small town and cooler clime lured them to Silver City, a place that had already earned a spot in Silke's heart when she had ridden long-distance horseback trail rides there a few years earlier. They bought a comfortable ranchette on the forested edge of the Gila Wilderness. There, Silke could have more room for horses, emus, turkeys, and dogs, and Richard could write with the windows open in the summer and grow desert and tropical plants year-round in his new custom greenhouse. Silke taught kindergarten in the town school and early childhood education classes at Western New Mexico University; Richard joined a local hiking club led by botanists and wrote a book with them about southwestern New Mexico trees.

Richard died on the Blue Moon of Halloween, October 31, 2020, a Saturday in Silver City, after three months of treatment for cancer and a week-long hospital stay in El Paso, Texas. Decades earlier he had survived prostate cancer. He was eighty-six. On the phone he had begged at least two friends to come kidnap him from his hospital bed in El Paso and take him home. It was an impossible request—he required regular pain medicines and constant attention, and it was during the height of a COVID epidemic, so hospitals and urgent care facilities along the way were full and could not take him if he worsened during the trip. But, following a surgery, he was able to ride back to Silver City in an ambulance. Even that had several delays, a driver who lacked

Richard with Apache redgrass at his home in Silver City, 2016. Photo by Bill Steen.

a license to cross state lines, then a snowstorm. Finally, Richard arrived home and felt comforted to pet his dogs and gaze at the forest he knew so well. He stopped breathing at 3:00 a.m., with his wife Silke by his side. Well-attended memorials were held at Silver City, New Mexico, in July 2021; the University of Arizona Herbarium garden in Tucson in April 2022; and at the Estero Soldado visitor center in San Carlos, Sonora, Mexico, along with the release of his book about the flora of that region in February 2024.

Friends and acquaintances will hold favorite memories of Richard: how he introduced them to a special plant or brought them a smile or seeds to sow. One such memory was his response to a holiday greeting card that showed two friends feeding a giraffe at the San Diego Zoo. Richard replied in his signature email style:

> all the best to you—at first i thought i was seeing you both in Africa but i see Washingtonia robusta and Phoenix canariensis in the background, so you must be visiting your long necked friend at the San Diego Zoo wildlife park.

When i was about 12 years old my dad bought me a membership in the San Diego Zoological Society and when my parents sold the big house we grew up in, my hundred or so cycad species / large and small / went to the San Diego Zoo and some are still there but they don't seem to remember me.[54]

*They* may not, but we sure do.

54. Richard Felger email to Bill Broyles and Joan Scott, December 23, 2016.

PART IV

# RICHARD S. FELGER'S BIBLIOGRAPHY

Compiled by Bill Broyles and Benjamin T. Wilder

## BOOKS


Felger, Richard Stephen. 2026. *The Botanical Odyssey of Richard S. Felger*. Edited by Benjamin T. Wilder, Bill Broyles, and Thomas Bowen. Tucson: University of Arizona Press.

Felger, Richard Stephen, and Felipe Molina. 2024. *Plants and Animals in the Yoeme World: Ethnoecology of the Yaquis of Sonora and Arizona*. Desert Institute Press.

Felger, Richard Stephen, Susan Davis Carnahan, and José Jesús Sánchez-Escalante. 2023. *The Desert Edge: Flora of the Guaymas–Yaqui Region of Sonora, Mexico*. 2 volumes. Fort Worth: BRIT Press.

Felger, Richard Stephen, James Thomas Verrier, Kelly Kindscher, and Xavier Raj Herbst Khera. 2021. *Field Guide to the Trees of the Gila Region of New Mexico*. Albuquerque: University of New Mexico Press.

Felger, Richard Stephen, and Benjamin T. Wilder, in collaboration with Humberto Romero-Morales. 2012. *Plant Life of a Desert Archipelago: Flora of the Sonoran Islands in the Gulf of California*. Tucson: University of Arizona Press.

Felger, Richard Stephen, and Bill Broyles, eds. 2007. *Dry Borders: Great Natural Reserves of the Sonoran Desert*. Salt Lake City: University of Utah Press.

Felger, Richard Stephen, with photographs by Patricio Robles Gil and George Huey. 2007. *The Secret Life of Cacti*. Agrupación Sierra Madre, Mexico City (also published in Spanish).

Cartron, Jean–Luc E., Gerardo Ceballos, and Richard Stephen Felger, eds. 2005. *Biodiversity, Ecosystems, and Conservation in Northern Mexico*. New York: Oxford University Press.

Felger, Richard Stephen, Matthew Brian Johnson, and Michael Francis Wilson. 2001. *Trees of Sonora, Mexico*. New York: Oxford University Press.

Felger, Richard Stephen. 2000. *Flora of the Gran Desierto and Río Colorado of Northwestern Mexico*. Tucson: University of Arizona Press.

Aguilar, Rafaela Paredes, Thomas R. Van Devender, and Richard Stephen Felger, in collaboration with Gary Paul Nabhan and Ana Lilia Reina Guerrero. 2000. *Las Cactáceas de Sonora: Su Diversidad, Uso y Conservación*. Tucson: Arizona-Sonora Desert Museum Press.

Felger, Richard Stephen, and Mary Beck Moser. (1985) 1991. *People of the Desert and Sea: Ethnobotany of the Seri Indians*. Tucson: University of Arizona Press.

Braniff C., Beatriz, and Richard Stephen Felger, eds. (1976) 1994. *Sonora, Antropología del Desierto*. Colección Cientifica Diversa 27. Mexico City: Instituto Nacional de Antropología e Historia. Reprint, Hermosillo, Mexico: Instituto Nacional de Antroplogía e Historia, Sonora. Conference papers and proceedings.

Felger, Richard Stephen. (1974) 2020. *Dark Horses and Little Turtles, and Other Poems of the Anthropocene*. Guadalupita, New Mexico: Nail Press, in a limited edition of 200 copies and as issue 1 of *Time and Tooth Review*, John Brandi, editor. New edition, Tucson: Polytropos Press, Greg McNamee, editor.

## INTERVIEWS, WEBSITES, AND VIDEOS

Richard Felger website: Desert Food Plants and More / The Work of Richard Stephen Felger (1934–2020), www.desertfoodplants.org.

Blystone, Peter. 2022. *Remembering Richard Felger: A Celebration of His Life and Legacy, Pax et Prosopis (Peace and Mesquite)*. University of Arizona Herbarium, Tucson, April 16. Video by Peter Blystone. https:/youtu.be/Pe9fJur667c?si=dbBR58wNSQbF4q0S.

Blystone, Peter. 2021. *Celebrating the Life of Dr. Richard S. Felger 1934–2020*. Video by Peter Blystone, uploaded November 3. https://www.youtube.com/watch?v=6mAraLozYMA.

Blystone, Peter. 2019. *Ethnobiology with Dr. Richard Felger (1934–2020). The Desert Was Both His Teacher and His Poetry*. Video by Peter Blystone, uploaded May 13. https://www.youtube.com/watch?v=dhnqINMB87U.

Ingram, Helen, and Gwynn Laird Benner. 2017. *Sonoran Desert Voices: Richard Felger, Ethnobotanist and Conservationist, University of Arizona*. Video by Peter Blystone, uploaded April 30. https:/youtu.be/XWOe174gkO0?si=wTuWgFy8C3R0TBt6.

## SCIENTIFIC PAPERS, BOOK CHAPTERS, AND POPULAR ARTICLES

Wilder, B. T., R. S. Felger, and E. Ezcurra. 2019. "Controls of Plant Diversity and Composition on a Desert Archipelago." *PeerJ* 7:e7286. https://doi.org/10.7717/peerj.7286.

Felger, R. S., S. D. Carnahan, and J. J. Sánchez-Escalante. 2017. "Oasis at the Desert Edge: Flora of Cañón del Nacapule, Sonora, Mexico." *Proceedings of the Desert Laboratory*, Contribution No. 1, 1–220.

Felger, R. S., S. D. Carnahan, and J. J. Sánchez-Escalante. 2017. "The Desert Edge—Flora of the Guaymas Region of Sonora, Mexico: The Checklist." *Desert Plants* 33 (1): 19–36.

Felger, R. S., S. D. Carnahan, J. A. Hawkins, and J. Verrier. 2017. "New Combinations for Sonoran Desert Plants." *Phytoneuron* 2017 (48): 1–6.

Felger, R. S., and S. Rutman. 2016. "Ajo Peak to Tinajas Altas: Flora of Southwestern Arizona, Part 21, Eudicots: Asteraceae." *Phytoneuron* 2016 (77): 1–66.

Felger, R. S., and S. Rutman. 2016. "Ajo Peak to Tinajas Altas: Flora of southwestern Arizona, Part 20, Eudicots: Solanaceae—Zygophyllaceae." *Phytoneuron* 2016 (52): 1–163.

Felger, R. S., and S. Rutman. 2016. "Ajo Peak to Tinajas Altas: Flora of Southwestern Arizona, Part 19, Eudicots: Polygalaceae to Simmondsiaceae." *Phytoneuron* 2016 (47): 1–71.

Felger, R. S., S. Rutman, and J. M. Porter. 2016. "Ajo Peak to Tinajas Altas: Flora of Southwestern Arizona, Part 18, Eudicots: Polemoniaceae–Phlox family." *Phytoneuron* 2016 (35): 1–24.

Felger, R. S., and S. Rutman. 2016. "Ajo Peak to Tinajas Altas: A Flora of Southwestern Arizona, Part 17, Eudicots: Nyctaginaceae to Plumbaginaceae." *Phytoneuron* 2016 (34): 1–77.

Felger, R. S., S. Rutman, C. J. S. Davis, and R. Lindley. 2015. "Ajo Peak to Tinajas Altas: A Flora of Southwestern Arizona, Part 16, Eudicots: Malpighiaceae to Moraceae." *Phytoneuron* 2015 (60): 1–54.

Felger, R. S., and S. Rutman. 2015. "Ajo Peak to Tinajas Altas: A Flora in Southwestern Arizona, Part 15, Eudicots: Fagaceae to Lythraceae." *Phytoneuron* 2015 (59): 1–53.

Felger, R. S., and S. Rutman. 2015. "Ajo Peak to Tinajas Altas: A Flora of Southwestern Arizona, Part 14, Eudicots: Fabaceae—Legume Family." *Phytoneuron* 2015 (58): 1–83.

Felger, R. S., S. Rutman, and N. C. Taylor. 2015. "Ajo Peak to Tinajas Altas: A Flora of Southwestern Arizona, Part 13, Eudicots: Euphorbiaceae." *Phytoneuron* 2015 (26): 1–65.

Felger, R. S., S. Rutman, and J. Malusa. 2015. "Ajo Peak to Tinajas Altas: A Flora of Southwestern Arizona, Part 12, Eudicots: Campanulaceae to Cucurbitaceae." *Phytoneuron* 2015 (21): 1–39.

Felger, R. S., S. Rutman, A. S. Salywon, and J. Malusa. 2015. "Ajo Peak to Tinajas Altas: A Flora of Southwestern Arizona, Part 11, Eudicots: Brassicaceae and Burseraceae." *Phytoneuron* 2015 (6): 1–48.

Felger, R. S., S. Rutman, C. M. Guilliams, and J. Malusa. 2015. "Ajo Peak to Tinajas Altas: A Flora of Southwestern Arizona, Part 10, Eudicots: Berberidaceae, Bignoniaceae, and Boraginaceae." *Phytoneuron* 2015 (1): 1–60.

Felger, R. S., S. Rutman, M. Costea, D. F. Austin, and J. Malusa. 2015. "Ajo Peak to Tinajas Altas: A Flora of Southwestern Arizona, Part 9, Eudicots: Convolvulaceae—Morning Glory Family." *Phytoneuron* 2015 (2): 1–22.

Felger, R. S., S. Rutman, and J. Malusa. 2014. "Ajo Peak to Tinajas Altas: A Flora of Southwestern Arizona, Part 8, Eudicots: Acanthaceae—Apocynaceae." *Phytoneuron* 2014 (85): 1–74.

Felger, R. S., S. Rutman, M. A. Baker, and J. Malusa. 2014. "Ajo Peak to Tinajas Altas: A Flora in Southwestern Arizona, Part 7, Eudicots: Cactaceae—Cactus Family." *Phytoneuron* 2014 (69): 1–95.

Felger, R. S. 2014. Foreword to *Shells on a Desert Shore: Mollusks in the Seri World*, by Cathy Moser Marlett. Tucson: University of Arizona Press.

Velarde, E., B. T. Wilder, R. S. Felger, and E. Ezcurra. 2014. "Floristic Diversity and Dynamics of Isla Rasa, Gulf of California, Mexico—A Globally Important Seabird Site." *Botanical Sciences* 92:89–101.

Felger, R. S., S. Rutman, and J. Malusa. 2014. "Ajo Peak to Tinajas Altas: A Flora of Southwestern Arizona, Part 6, Poaceae—Grass Family." *Phytoneuron* 2014 (35): 1–139.

Vanderplank, S., E. Ezcurra, J. Delgadillo, R. Felger, and L. A. McDade. 2014. "Conservation Challenges in a Threatened Hotspot: Agriculture and Plant Biodiversity Losses in Baja California, Mexico." *Biodiversity and Conservation* 23:2173–82.

Malusa, J., K. Reichardt, and R. S. Felger. 2013. "Giant Sandbur (*Cenchrus palmeri*, Poaceae) New for Arizona and the United States." *Phytoneuron* 2013 (91): 1–5.

Felger, R. S., S. Rutman, and J. Malusa. 2013. "Ajo Peak to Tinajas Altas: A Flora of Southwestern Arizona, Part 5, Monocots Except Grasses." *Phytoneuron* 2013 (76): 1–59.

Felger, R. S., S. Rutman, J. Malusa, and T. R. Van Devender. 2013. "Ajo Peak to Tinajas Altas: A Flora of Southwestern Arizona, Part 4, Angiosperms: Magnoliids." *Phytoneuron* 2013 (38): 1–9.

Felger, R. S., S. Rutman, J. Malusa, and T. R. Van Devender. 2013. "Ajo Peak to Tinajas Altas: A Flora of Southwestern Arizona, Part 3, Ferns, Lycopods, and Gymnosperms." *Phytoneuron* 2013 (37): 1–46.

Felger, R. S., S. Rutman, J. Malusa, and T. R. Van Devender. 2013. "Ajo Peak to Tinajas Altas: A Flora of Southwestern Arizona, Part 2, The Checklist." *Phytoneuron* 2013 (27): 1–30.

Felger, R. S., S. Rutman, J. Malusa, and T. R. Van Devender. 2013. "Ajo Peak to Tinajas Altas: A Flora of Southwestern Arizona—An Introduction." *Phytoneuron* 2013 (5): 1–40.

Glenn, E. P., T. Anday, R. Chaturvedi, R. Martinez-Garcia, S. Pearlstein, D. Soliz, S. G. Nelson, and R. S. Felger. 2013. "Three Halophytes for Saline-Water Agriculture: An Oilseed, a Forage and a Grain Crop." *Environmental and Experimental Botany* 92:110–21.

Felger, R. S., D. F. Austin, T. R. Van Devender, J. J. Sánchez-Escalante, and M. Costea. 2012. "Convolvulaceae of Sonora, Mexico, *I. Convolvulus*, *Cressa*, *Dichondra*, *Evolvulus*, *Ipomoea*, *Jacquemontia*, *Merremia*, and *Operculina*." *Journal of the Botanical Research Institute of Texas* 6 (2): 459–527.

Costea, M., R. S. Felger, D. F., Austin, T. R. Van Devender, and J. J. Sánchez-Escalante. 2012. "Convolvulaceae of Sonora, Mexico, II, *Cuscuta*." *Journal of the Botanical Research Institute of Texas* 6 (2): 529–50.

Norris, W., and R. S. Felger, eds. 2012. "Proceedings of the Third Natural History of the Gila Symposium, October 14–16, 2010, Western New Mexico University Silver City, New Mexico." Special issue, *New Mexico Botanist* 3:1–68. http://sites.google.com/view/gilasymposium/published-proceedings.

Pearlstein, S. L., R. S. Felger, E. P. Glenn, J. Harrington, K. A. Al-Ghanem, and S. G. Nelson. 2012. "Nipa (*Distichlis palmeri*): A Perennial Grain Crop for Saltwater Irrigation." *Journal of Arid Environments* 82:60–70.

Felger, R. S., S. Rutman, T. R. Van Devender, and S. M. Buckley. 2012. "Checklist of Vascular Plants of Organ Pipe Cactus National Monument, Cabeza Prieta National Wildlife Refuge, and Tinajas Altas, Arizona." *Canotia* 8:1–53.

Felger, R. S., T. R. Van Devender, B. Broyles, and J. Malusa. 2012. "Flora of Tinajas Altas, Arizona—A Century of Botanical Forays and Forty Thousand Years of Neotoma Chronicles." *Journal of the Botanical Research Institute of Texas* 6 (1): 157–257.

Gallo-Reynoso, J. P., R. S. Felger, and B. T. Wilder. 2012. "Near Colonization of a Desert Island by a Tropical Bird: Military Macaw (*Ara militaris*) at Isla San Pedro Nolasco, Gulf of California." *Southwestern Naturalist* 57 (4): 459–62.

Felger, R. S., and T. R. Van Devender. 2012. "Vascular Plants of the Tinajas Altas Region, Arizona." In *Last Water on the Devil's Highway: A Cultural and Natural History of Tinajas Altas*, edited by B. Broyles, G. H. Hartmann, T. E Sheridan, G. P. Nabhan, and M. C. Thurtle, 213–32. Tucson: University of Arizona Press.

Felger, R. S., B. T. Wilder, and J. P. Gallo-Reynoso. 2011. "Floristic Diversity and Long-term Vegetation Dynamics of Isla San Pedro Nolasco, Gulf of California, Mexico." *Proceedings of the San Diego Society of Natural History* 43:1–42.

Cain, J. W., III, B. D. Jansen, R. S., Felger, and P. R. Krausman. 2010. "Scallopleaf Sage (*Salvia vaseyi*, Lamiaceae) Discovered in Arizona." *Journal of the Botanical Research Institute of Texas* 4 (2): 755–60.

Felger, R. S., and K. Kindscher. 2010. "Trees of the Gila Forest Region, New Mexico." In Proceedings of the Second Natural History of the Gila Symposium, October 16–18, 2008, Silver City, New Mexico, edited by W. Norris, Felger, and K. Kindscher. Special issue, *New Mexico Botanist* 2: 38–66.

Wilder, B. T., and R. S. Felger. 2010. "Dwarf Giants, Guano, and Isolation: Vegetation and Floristic Diversity of Isla San Pedro Mártir, Gulf of California, Mexico." *Proceedings of the San Diego Society of Natural History* 42:1–24.

Martínez-Yrízar, A., R. S. Felger, and A. Búrquez. 2010. "Los Ecosistemas Terrestres de Sonora: Un Diverso Capital Natural." In *Diversidad Biologíca del Estado de Sonora*, edited by F. Molina-Freanar and T. R. Van Devender, 129–56. Mexico City: Universidad Nacional Autónoma de México.

Van Devender, T. R., R. S. Felger, F. Molina-Freanar, M. Fishbein, J. J. Sánchez-Escalante, and A. L. Reina-Guerrero. 2010. "Biodiversidad de Plantas Vasculares." In *Diversidad Biologíca del Estado de Sonora*, edited by F. Molina-Freanar and T. R. Van Devender, 229–62. Mexico City: Universidad Nacional Autónoma de México.

Van Devender, T. R., R. S. Felger, A. L. Reina-G., and J. J. Sánchez-Escalante. 2009. "Non-native and Invasive Plants in Sonora, Mexico." In *Invasive Plants on the Move: Controlling Them in North America*, edited by T. R. Van Devender, F. J. Espinosa-García, B. L. Harper-Lore, and T. Hubbard, 85–124. Proceedings of Weeds Across Borders 2006 Conference, Hermosillo, Sonora, May 25–28, 2006. Tucson: Arizona-Sonora Desert Museum.

Austin, D. F., and R. S. Felger. 2008. "Sichuan Peppers and the Etymology of *Fagara* (Rutaceae)." *Economic Botany* 62 (4): 567–73.

Wilder, B. T., R. S. Felger, and H. Romero-Morales. 2008. "Succulent Plant Diversity of the Sonoran Islands, Gulf of California, Mexico." *Haseltonia* 14:127–60.

Wilder, B. T., R. S. Felger, T. R. Van Devender, and H. Romero-Morales. 2008. *Canotia holacantha* on Isla Tiburón, Gulf of California, Mexico. *Canotia* 4 (1): 1–7.

Van Devender, T. R., R. S. Felger, and M. A. Dimmitt. 2007. "Sonoran Desert Weed Accounts." In *Roadside Weed Management*, edited by B. L. Harper-Lore, M. Johnson, and M. W. Skinner, 28–29. Washington, D.C.: U.S. Department of Transportation, Federal Highway Administration.

Van Devender, T. R., R. S. Felger, and M. A. Dimmitt. 2007. "Buffelgrass (*Pennisetum ciliare*)." In *Roadside Weed Management*, edited by B. L. Harper-Lore, M. Johnson, and M. W. Skinner, 30. Washington, D.C.: U.S. Department of Transportation, Federal Highway Administration.

Wilder, B. T., R. S. Felger, H. Romero-Morales, and A. Quijada-Mascareñas. 2007. "New Plant Discoveries for the Sonoran Islands, Gulf of California, Mexico." *Journal of the Botanical Research Institute of Texas* 1: 1203–27.

Felger, R. S., B. Broyles, M. F. Wilson, G. P. Nabhan, and D. S. Turner. 2007. "Six Grand Reserves, One Grand Desert." In *Dry Borders: Great Natural Reserves of the Sonoran Desert*, edited by R. S. Felger and B. Broyles, 3–26. Salt Lake City: University of Utah Press.

Broyles, B., A. G. Rankin, and R. S. Felger. 2007. "Native Peoples of the Dry Borders Region." In *Dry Borders: Great Natural Reserves of the Sonoran Desert*, edited by R. S. Felger and B. Broyles, 128–46. Salt Lake City: University of Utah Press.

Felger, R. S. 2007. "Living Resources at the Center of the Sonoran Desert: Native American Plant and Animal Utilization." In *Dry Borders: Great Natural Reserves of the Sonoran Desert*, edited by R. S. Felger and B. Broyles, 147–92. Salt Lake City: University of Utah Press.

Felger, R. S., S. Rutman, M. F. Wilson, and K. Mauz. 2007. Botanical Diversity of Southwestern Arizona and Northwestern Sonora. In *Dry Borders: Great Natural Reserves of the Sonoran Desert*, edited by R. S. Felger and B. Broyles, 202–71. Salt Lake City: University of Utah Press.

Broyles, B., L. Evans, R. S. Felger, and G. P. Nabhan. 2007. "Our Grand Desert: A Gazetteer." In *Dry Borders: Great Natural Reserves of the Sonoran Desert*, edited by R. S. Felger and B. Broyles, 509–607. Salt Lake City: University of Utah Press.

Felger, R. S., and D. F. Austin. 2005. "*Ipomoea seaania*, a New Species of Convolvulaceae from Sonora, Mexico." *SIDA, Contributions to Botany* 21:1293–1304.

Stoleson, S. H., R. S. Felger, G. Ceballos, C. Raish, M. F. Wilson, and A. Búrquez. 2005. "Recent History of Natural Resource Use and Population Growth in Northern Mexico." In *Biodiversity, Ecosystems, and Conservation in Northern Mexico*, edited by J-L. E. Cartron, G. Ceballos, and R. S. Felger, 52–86. New York: Oxford University Press.

Whitmore, R. C., R. C. Brusca, J. L. de la Luz, P. González-Zamorano, R. Mendoza-Salgado, E. S. Amador-Silva, G. Holguin, F. Galván-Magaña, P. A. Hastings, J.-L. E. Cartron, R. S. Felger, J. A. Seminoff, and C. C. McIvor. 2005. "The Ecological Importance of Mangroves in Baja California Sur: Conservation Implications for an Endangered Ecosystem." In *Biodiversity, Ecosystems, and Conservation in Northern Mexico*, edited by J-L. E. Cartron, G. Ceballos, and R. S. Felger, 298–333. New York: Oxford University Press.

Felger, R. S., W. J. Nichols, and J. A. Seminoff. 2005. "Sea Turtles in Northwestern Mexico: Conservation, Ethnobiology, and Desperation." In *Biodiversity, Ecosystems, and Con-*

*servation in Northern Mexico*, edited by J-L. E. Cartron, G. Ceballos, and R. S. Felger, 405–24. New York: Oxford University Press.

Felger, R. S., T. L. Burgess, S. Dorsi, J. R. Reeder, and T. R. Van Devender. 2005. "*Dichanthium* (Poaceae) New to Arizona: Open Door for a Potentially Invasive Species." *SIDA, Contributions to Botany* 21:1905–8.

Austin, D. F., R. S. Felger, and T. R. Van Devender. 2005. "Nomenclature of *Ipomoea arborescens* (Convolvulaceae) in Sonora, Mexico." *SIDA, Contributions to Botany* 21:1283–92.

Bowen, T., R. S. Felger, and R. J. Hills. 2004. "Chollas, Circles, and Seris: Did Seri Indians Plant Cactus at Circle 6?" *Desert Plants* 20 (2): 26–35.

Borchert, R., S. A. Meyer, R. S. Felger, and L. Porter-Bolland. 2004. "Environmental Control of Flowering Periodicity in Costa Rican and Mexican Tropical Dry Forests." *Global Ecology and Biogeography* 13:409–35.

Felger, R. S., W. J. Nichols, A. Aguierre, H. Aridjis, L. Bourillón-Moreno, R. Brusca, J. Compoy, A. Cantu, J.-L. Cartron, G. D. Daneman, H. D'Antoni, P. K. Dayton, S. A. Diehn, L. Gerber, J. L. León de la Luz, C. Mártinez del Rio, R. A. Medellín, E. Palacios Castro, and E. Sala. 2004. "The Escalera Ecológica: A Staircase of Reserves through Southwestern Arizona and the Gulf of California Region (abstract)." In *Proceedings, Gulf of California Conference 2004*, 71. Tucson: Arizona-Sonora Desert Museum.

Felger R. S. 2004. "Seed Plants." In *A Seashore Guide to the Northern Gulf of California*, edited by Richard C. Brusca, Erin Kimrey, and Wendy Moore, 147–63. Tucson: Arizona-Sonora Desert Museum.

Felger, R. S., D. S. Turner, and M. F. Wilson. 2003. "Flora and Vegetation of the Mohawk Dunes, Arizona." *SIDA, Contributions to Botany* 20:1153–85.

Felger, R. S. 2003. "Sea Turtles in Northwestern Mexico: Conservation, Ethnobiology, and Desperation." In *Proceedings of the Twenty-Second Annual Symposium on Sea Turtle Biology and Conservation*, edited by J. A. Seminoff, 82–83. NOAA Technical Memorandum. National Marine Fisheries Service-SEFSC-503.

Yetman, D., and R. S. Felger. 2002. "Ethnoflora of the Guarijíos." In *Guarijíos of the Sierra Madre: The Hidden People of Northwestern Mexico*, edited by D. Yetman, 174–230. Albuquerque: University of New Mexico Press.

Felger, R. 2002. "Save the Desert, Prevent a Weed." Preface and introduction to *Invasive Plants of the Sonoran Desert: A Field Guide*, 4–6. Tucson, Ariz.: Sonoran Institute, Environmental Exchange, and National Fish and Wildlife Foundation (also published in Spanish).

Wilson, M. F., L. Leigh, and R. S. Felger. 2002. "Invasive Exotic Plants in the Sonoran Desert." In *Invasive Exotic Species in the Sonoran Region*, edited by B. Tellman, 81–90. Tucson: University of Arizona Press.

Nabhan, G. P., P. West, R. S. Felger, et al. 2002. "Naturalized Exotic Species in the Sonoran Region, Flora." In *Invasive Exotic Species in the Sonoran Region*, edited by B. Tellman, 348–55. Tucson: University of Arizona Press.

Felger, R. S. 2002. "Sinaloa Shootout." In *Backcountry Pilot: Flying Adventures with Ike Russell*, edited by T. Bowen, 2–14. Tucson: University of Arizona Press.

Broyles, B., R. S. Felger, and C. Bowden. 2001. "The Sonoran Corridor." In *The Gulf of California, A World Apart*, edited by P. Robles Gil, E. Ezcurra, and E. Mellink, 245–65. Mexico City: Agrupación Sierra Madre.

Felger, R. S. 2001. "Coastal Wetlands." In *The Gulf of California, A World Apart*, edited by P. Robles Gil, E. Ezcurra, and E. Mellink, 159–81. Mexico City: Agrupación Sierra Madre.

Reina-Guerrero, A. L., T. R. Van Devender, D. Yetman, and R. S. Felger. 2000. "Usos Etnobótanicos de las Cactáceas en Sonora, México." In *II Simposium Internacional sobre la Utilización y Aprovechamiento de la Flora Silvestre de Zonas Aridas*, edited by D. Vásquez del Castillo, M. Ortegas N., R. A. Corella B., and R. A. Castillo G., 191–202. Hermosillo: Departmento de Investigaciones Científicas y Technologicas de la Universidad de Sonora.

Felger R. S., and D. Yetman. 2000. "Roasting the Hechtia Out of It: The Use of *Hechtia montana* (Bromeliaceae) as a Food Plant in Sonora, México." *Economic Botany* 54:229–33.

Turner, D. S., R. S. Felger, K. Mauz, C. Funicelli, T. Van Devender, and J. Malusa. 2000. "Biological Resources of the Proposed Sonoran Desert National Monument, Arizona." Tucson, Ariz.: Sonoran Desert National Park Project.

Felger, R. S. 2000. "The Seris and the Guy Who Cuts the Tops Off Plants." In "Seri Hands," special issue, *Journal of the Southwest* 42:521–43.

Felger, R. S. 1999. "The Flora of Cañón del Nacapule: A Desert-Bounded Tropical Canyon near Guaymas, Sonora, Mexico." *Proceedings of the San Diego Society of Natural History* 35:1–42.

Felger, R. S., and E. Joyal. 1999. "The Palms (Arecaceae) of Sonora, Mexico." *Aliso* 18:1–18.

Búrquez, A., A. Martínez-Yrízar, R. S. Felger, and D. Yetman. 1999. "Vegetation and Habitat Diversity at the Southern Edge of the Sonoran Desert." In *Ecology of Sonoran Desert Plants and Plant Communities*, edited by R. H. Robichaux, 36–67. Tucson: University of Arizona Press.

Felger, R. S. 1998. "Cupressaceae—Cypress Family." In *Gentry's Río Mayo Plants*, edited by P. S. Martin et al., 191–92. Tucson: University of Arizona Press.

Felger, R. S., and M. A. Dimmitt. 1998. "Bromeliaceae." In *Gentry's Río Mayo Plants*, edited by P. S. Martin et al., 475–77. Tucson: University of Arizona Press.

Felger, R. S., and M. A. Dimmitt. 1998. "Orchidaceae." In *Gentry's Río Mayo Plants*, edited by P. S. Martin et al., 492–98. Tucson: University of Arizona Press.

Felger, R. S., and E. Joyal. 1998. "Arecaceae." In *Gentry's Río Mayo Plants*, edited by P. S. Martin et al., 474. Tucson: University of Arizona Press.

Steinmann, V. W., and R. S. Felger. 1997. "*Croton yecorensis* (Euphorbiaceae), a New Species from Northwestern Mexico." *Novon* 8:207–9.

Steinmann, V. W., and R. S. Felger. 1997. "The Euphorbiaceae of Sonora, Mexico." *Aliso* 16:1–71.

Felger, R. S., G. P. Nabhan, and R. Bye. 1997. "The Apachian/Madrean Region of Southwestern North America as a Center of Plant Diversity." In *Centres of Plant Diversity: A Guide and Strategy for Their Conservation*, vol. 3, *The Americas*, edited by S. D. Davis et al., 172–80. Cambridge: World Wide Fund for Nature and International Union for Conservation and Nature.

Felger, R. S., and B. Broyles, eds. 1997. "Dry Borders." Special issue, *Journal of the Southwest* 39:303–860.

Felger, R. S. 1997. "A Botanist's View of the Center of the Universe." *Journal of the Southwest* 39:399–410.

Felger R. S., M. F. Wilson, B. Broyles, and G. P. Nabhan. 1997. "The Binational Sonoran Desert Biosphere Network and Its Plant Life." *Journal of the Southwest* 39:411–560.

Broyles, B., R. S. Felger, G. P. Nabhan, and L. Evans. 1997. "Our Grand Desert: A Gazetteer for Northwestern Sonora, Southwestern Arizona, and Northeastern Baja California." *Journal of the Southwest* 39:703–855.

Felger, R. S., and J. Henrickson. 1997. "Convergent Adaptive Morphology of a Sonoran Desert Cactus (*Peniocereus striatus*) and an African Spurge (*Euphorbia cryptospinosa*)." *Haseltonia* 5:77–85.

Van Devender, T. R., R. S. Felger, and A. Búrquez. 1997. "Exotic Plants in Sonora, Mexico." In Kelley, Wagner, and Warner, *Proceedings California Exotic Pest Plant Council Symposium*, vol. 3, 10–15.

Glenn, E. P., C. Lee, R. S. Felger, and S. Zengel. 1996. "Effects of Water Management on the Wetlands of the Colorado River Delta, Mexico." *Conservation Biology* 10:1175–86.

Friedman, S. L., T. R. Van Devender, A. L. Reina-Guerrero, D. A. Yetman, V. W. Steinmann, A. C. Sanders, P. D. Jenkins, S. A. Meyer, R. S. Felger, and R. A. López E. 1996. "Noteworthy Collections: Sonora." *Madroño* 43:532–38.

Steinmann, V. W., and R. S. Felger. 1995. "New Combinations for *Euphorbia* and *Ditaxis* (Euphorbiaceae) in Northwestern Mexico and Southwestern United States." *Madroño* 42:455–57.

Felger, R. S., and M. F. Wilson. 1995. "Northern Sierra Madre Occidental and Its Apachian Outliers: A Neglected Center of Biodiversity." In *Biodiversity and Management of the Madrean Archipelago: The Sky Islands of Southwestern United States and Northern Mexico*, edited by L. DeBano et al., 36–59. Gen. Tech. Rep. RM-GTR-264. Fort Collins, Colo.: U.S. Department of Agriculture, Forest Service, Rocky Mountain Forest Range Experiment Station.

Felger, R. S., and M. B. Johnson. 1995. "Trees of the Northern Sierra Madre Occidental and Sky Islands of Southwestern North America." In *Biodiversity and Management of the Madrean Archipelago: The Sky Islands of Southwestern United States and Northern Mexico*, edited by L. DeBano et al., 71–83. Gen. Tech. Rep. RM-GTR-264. Fort Collins, Colo.: U.S. Department of Agriculture, Forest Service, Rocky Mountain Forest Range Experiment Station.

Fishbein, M., R. S. Felger, and F. Garza. 1995. "Another Jewel in the Crown: A Report on the Flora of the Sierra De Los Ajo, Sonora, Mexico." In *Biodiversity and Management of the Madrean Archipelago: The Sky Islands of Southwestern United States and Northern Mexico*, edited by L. DeBano et al., 126–34. Gen. Rech. Rep. RM-GTR-264. Fort Collins, Colo.: U.S. Department of Agriculture, Forest Service, Rocky Mountain Forest Range Experiment Station.

Felger, R. S., and A. Búrquez. 1995. "Noteworthy Collections." *Madroño* 42:517–18.

Fishbein, M., and R. S. Felger. 1995. "Noteworthy Collections." *Madroño* 42:409–10.

Zengel, S. A., V. Meretsky, E. P. Glenn, R. S. Felger, and D. Ortiz. 1994. "Cienega de Santa Clara, a Remnant Wetland in the Rio Colorado Delta (Mexico): Vegetation Distribution and the Effects of Water Flow Reduction." *Ecological Engineering* 4:19–36.

Felger, R. S., A. Harlan, V. W. Steinmann, and F. W. Telewski. 1993. "*Cryptantha ganderi* I. M. Johnston (Boraginaceae): New for Arizona." *Madroño* 40:268.

Felger, R. S. 1993. "*Mirabilis tenuiloba* S. Watson (Nyctaginaceae): New for Arizona." *Madroño* 40:178.

Felger, R. S. 1992. "Synopsis of the Vascular Plants of Northwestern Sonora, Mexico." *Ecologica* 2 (2): 11–44.

Felger, R. S. 1992. "Reflections on a Desert Legume Trinity." *Aridus* 4 (4): 1–5, 7.

Glenn, E. P., R. S. Felger, A. Búrquez, and D. S. Turner. 1992. "Cienega de Santa Clara: Endangered Wetland in the Colorado River Delta, Sonora, Mexico." *Natural Resources Journal* 32 (4): 17–824.

Felger, R. S. 1992. "Recursos Bióticos del Desierto de Sonora." In *Ecología, Recursos Naturales y Medio Ambiente en Sonora*, edited by J. L. Moreno, 91–97. Hermosillo: Secretaría de Infraestructura Urbana y Ecología and El Colegio de Sonora.

Felger, R. S., P. L. Warren, L. S. Anderson, and G. P. Nabhan. 1992. "Vascular Plants of a Desert Oasis: Flora and Ethnobotany of Quitobaquito, Organ Pipe Cactus National Monument, Arizona." In *Proceedings of the San Diego Society of Natural History* 8:1–39.

Pinkava, D. J., M. A. Baker, R. A. Johnson, N. Trushell, G. A. Ruffner, R. S. Felger, and R. K. Van Devender. 1992. "Additions, Notes, and Chromosome Numbers for the Vascular Flora of Organ Pipe Cactus National Monument, Arizona." *Journal of the Arizona-Nevada Academy of Science* 24–25:13–18.

Felger, R. S. 1991. "*Senecio pinacatensis* (Asteraceae), a New Species from the Pinacate Region of Sonora, Mexico." *Phytologia* 71:326–32.

Glenn, E. P., R. S. Felger, and J. A. Búrquez. 1991. "Oasis en el Delta del Río Colorado, Ciénega de Santa Clara: Vestigios de un Oasis de Humedal." *Noticias del CEDO* (Centro Intercultural de Estudios de Desiertos y Oceanos, Mexico) 4 (1): 14–15, 19–29, 31–32. (Reprinted 1992 in *Arizona Riparian Council Newsletter*).

Van Devender, T. R., T. L. Burgess, R. S. Felger, and R. M. Turner. 1990. "Holocene Vegetation of the Hornaday Mountains of Northwestern Sonora, Mexico." In *Proceedings of the San Diego Society of Natural History* 2:1–19.

Felger, R. S. 1990. "Global Diversity of New Food Crops." In *Proceedings: 1st International Conference on Promising Crops for Arid and Semiarid Lands*. Asunción, Paraguay.

Felger, R. S. 1990. "New Crop Potential Among Columnar Cacti." *Proceedings: 1st International Conference on Promising Crops for Arid and Semiarid Lands*. Asunción, Paraguay.

Felger, R. S. 1990. "Non-native Plants of Organ Pipe Cactus National Monument." Technical Report 31. Cooperative National Park Resource Studies Unit, University of Arizona, Tucson.

Felger, R. S. 1990. "Mesquite, a World Food Crop." *ARIDUS* 2 (1): 1–3.

Felger, R. S. 1990. "The Seri Indians and Their Herpetafauna." *Sonoran Herpetologist* 3 (5): 41–44.

Felger, R. S. 1990. "Los Indios Seri, Caguamas y el Desierto." *Noticias del CEDO* 3 (1): 21–24; *Los Indios Seri, el Mar y el Desierto* 3 (2): 20–23.

Reeder, J. R., and R. S. Felger. 1989. "The *Aristida californica-glabrata* Complex (Gramineae)." *Madroño* 36 (3): 187–97.

Felger, R. S., F. A. Molina, and M. B. Moser. 1989. "Cactus Songs (Seri and Yaqui)." *Plant Press* 13 (1): 8.

Felger, R. S. 1988. "*Pilostyles*, Strange Desert Parasite." *Plant Press* 12 (2): 6–7.

Ezcurra, E., R. S. Felger, A. Russell, and M. Equihua. 1988. "Fresh Water Islands in a Desert Sand Sea: The Hydrology, Flora, and Phytogeography of the Gran Desierto Oases of Northwestern Mexico." *Desert Plants* 9 (2): 35–44, 55–63.

Felger, R. S., and M. B. Moser. 1987. "Sea Turtles in Seri Indian Culture." *Environment Southwest*, San Diego Museum of Natural History no. 519, 18–21.

Felger, R. S. 1985. "New Arid Land Crops from Mexico: Three Examples for Immediate Development." In *Uso y Preservación de los Recursos Biológicas Marinos y de Zonas Aridas*, edited by J. L. Ochoa and J. Moreno-Lopez, 13–24. La Paz, Baja California Sur: Centro de Investigaciones Biológicas de Baja California Sur.

Nabhan, G. P., and R. S. Felger. 1985. "Wild Desert Relatives of Crops: Their Direct Use as Food." pp. In *Economic Uses of Arid Land Plants*, edited by G. Wickens, 19–33. Kew: Royal Botanic Gardens.

Felger, R. S., and M. B. Moser. 1984. "Use and Knowledge of Legumes (Mimosoideae) by the Seri Indians of Sonora, Mexico." *Bulletin of the International Group for the Study of Mimosoideae* 12:78–82.

Felger, R. S., M. B. Moser, and E. W. Moser. 1983. "The Desert Tortoise in Seri Indian Culture." In *Desert Tortoise Council, Proceedings of 1981 Symposium*, 113–20. Long Beach, Calif.: Desert Tortoise Council.

Cliffton, K., D. O. Cornejo, and R. S. Felger. 1982. "Sea Turtles of the Pacific Coast of Mexico." In *Biological Conservation of Sea Turtles*, edited by K. Bjorndal, 199–209. Washington, D.C.: Smithsonian Institution Press.

Cornejo, D. O., L. S. Leigh, R. S. Felger, and C. F. Hutchinson. 1982. "Utilization of Mesquite in the Sonoran Desert: Past and Future." In *Mesquite Utilization, Proceedings of the Symposium*, edited by H. W. Parker, 1–20. Lubbock: Texas Tech University.

Felger, R. S. 1982. "Review of: L. Benson and R. A. Darrow, *Trees and Shrubs of the Southwestern Deserts*, University of Arizona Press, Tucson, 1981." *Quarterly Review of Biology* 57:333–34.

Felger, R. S., and C. Mota-Urbana. 1982. "Halophytes: New Sources of Nutrition." In *Biosaline Research: A Look to the Future*, edited by A. San Pietro, 473–78. New York: Plenum.

Yensen, N. P., M. R. Fontes, E. P. Glenn, and R. S. Felger. 1981. "New Salt Tolerant Crops for the Sonoran Desert." *Desert Plants* 3 (3): 111–17.

Felger, R. S. 1981. "Future Harvests from Seeds of the Past." *Sonorensis* 3 (3): 6–7.

Felger, R. S. 1981. "Review of: I. L. Wiggins, *Flora of Baja California*, Stanford University Press, Stanford, 1980." *Taxon* 30: 533–34.

Felger, R. S., L. S. Leigh, S. L. Buchmann, D. O. Cornejo, M. A. Dimmitt, D. Johnson Gordon, C. Nagel, L. Ratener, and C. A. Stigers. 1981. "Inventorying the World's Arid Lands for New Crops: A Model from the Sonoran Desert." In *Proceedings of Arid Lands Resource Inventories Workshop*, edited by H. G. Lund, 106–16. USDA Forest Service General Technical Report W0–28.

Felger, R. S., M. B. Moser, and E. W. Moser. 1980. "Seagrasses in Seri Indian Culture." In *Handbook of Seagrass Biology: An Ecosystem Perspective*, edited by R. C. Phillips and C. P. McRoy, 260–76. New York: Garland STPM Press.

Felger, R. S. 1980. "Vegetation and Flora of the Gran Desierto, Sonora, Mexico." *Desert Plants* 2 (2): 87–114.

Felger, R. S., D. O. Cornejo, L. S. Leigh, and C. A. Stigers. 1979. "Semillas Magicas del Desierto: Un Modelo de Agricultura Apropiada del Desierto Sonorense." In *IV Simposio Sobre El Medio Ambiente del Golfo de California, Memoria*, 70–80. Mexico City: Instituto Nacional de Investigaciones Forestales.

Felger, R. S. 1979. "Ancient Crops for the 21st Century." In *New Agricultural Crops*, edited by G. Ritchie, 5–20. AAAS Selected Symposium 38. Boulder, Colo.: Westview.

Felger, R. S., and R. C. Wilkinson. 1978. "Feast of Life: Journal for the Protection of All Beings." *Co-evolutionary Quarterly* 19:83–90.

Felger, R. S., and G. P. Nabhan. 1978. "Agroecosystem Diversity: A Model from the Sonoran Desert." In *Social and Technological Management in Dry Lands*, edited by N. L. Gonzalez, 128–49. AAAS Selected Symposium 10. Boulder, Colo.: Westview.

Nabhan, G. P., and R. S. Felger. 1978. "Teparies in Southwestern North America." *Economic Botany* 32 (1): 2–19.

Felger, R. S. 1977. "Mesquite in Indian Cultures of Southwestern North America." In *Mesquite: Its Biology in Two Desert Scrub Ecosystems*, edited by B. B. Simpson, 150–76. Stroudsburg, Pa.: Dowden, Hutchinson, and Ross.

Sheridan, T. E., and R. S. Felger. 1977. "Indian Utilization of Eelgrass (*Zostera marina* L.) in Northwestern Mexico: The Spanish Colonial Record." *Kiva* 43:89–92.

R. Menninger, interview with R. S. Felger. 1977. "New Crops." *Co-Evolutionary Quarterly*, Fall, 76–79.

Nabhan, G. P., and R. S. Felger. 1977. "Ancient Crops for Desert Gardens." *Organic Gardening and Farming* 24 (2): 34–42.

Felger, R. S., and G. P. Nabhan. 1976. "Deceptive Barrenness." *Ceres* 9 (2): 34–39 (also published in Spanish and French).

Felger, R. S., and M. B. Moser. 1976. "Seri Indian Food Plants: Desert Subsistence Without Agriculture." *Ecology of Food and Nutrition* 5:13–27.

Felger, R. S. (1976) 1994. "Investigaciones Ecologicas en Sonora y Localidades Adyacentes en Sinaloa—Una Perspectiva." In B. Braniff and R. S. Felger, *Sonora, Antropología del Desierto*, 21–62.

Felger, R. S., K. Cliffton, and P. J. Regal. 1976. "Winter Dormancy in Sea Turtles: Independent Discovery and Exploitation in the Gulf of California by Two Local Cultures." *Science* 191:283–85.

Felger, R. S., and C. H. Lowe. 1976. *The Island and Coastal Vegetation and Flora of the Gulf of California, Mexico*. Contributions in Science 285. Natural History Museum of Los Angeles County.

Felger, R. S. 1976. "The Gulf of California: An Ethno-ecological Perspective." *Natural Resources Journal* 16 (3): 451–64.

Contributor with various authors. 1975. "El Golfo de California, un Recurso Mexicano de Importancia Internacional." *Supervivencia* 1 (4): 13–22.

Felger, R. S., and C. P. McRoy. 1975. "Seagrasses as Potential Food Plants." In *Seed-Bearing Halophytes as Food Plants*, edited by G. F. Somers, 62–74. Newark: University of Delaware.

Felger, R. S. 1975. "Nutritionally Significant New Crops for Arid Lands: A Model from the Sonoran Desert." In *Priorities in Child Nutrition in Developing Countries*, edited by J. Mayer and J. W. Dwyer, 373–403. New York: UNICEF.

Felger, R. S. 1975. "Review of: C. Niethammer, American Indian Food and Lore, Macmillan, New York, 1974." *Journal of Arizona History* 16:302–3.

Felger, R. S., and M. B. Moser. 1974. "Columnar Cacti in Seri Indian Culture." *Kiva* 39: 257–75.

Felger, R. S., and M. B. Moser. 1974. "Seri Indian Pharmacopoeia." *Economic Botany* 28 (4): 414–36.

Felger, R. S., and M. B. Moser. 1973. "Eelgrass (*Zostera marina* L.) in the Gulf of California: Discovery of Its Nutritional Value by the Seri Indians." *Science* 181:355–56.

Henrickson, J., and R. S. Felger. 1973. "Microanalysis and Identification of a Basket Fragment from Sonora, Mexico." *Kiva* 38 (3–4): 173–77.

Bunnell, S., and R. S. Felger. 1972. "The Sonoran Desert." *Clear Creek* 13:44–46.

Felger, R. S. 1972. "Stockholm Superconference." *Defenders of Wildlife* 47 (3): 247–49.

Felger, R. S. 1972. "Fog Desert." *Pacific Discovery* 25 (3): 14–18.

Felger, R. S., and M. B. Moser. 1971. "Seri Use of Mesquite (*Prosopis glandulosa* var. *torreyana*)." *Kiva* 37 (3–4): 53–60.

Felger, R. S. 1971. "*Magnolia* in Northwestern Mexico." *Journal of the Arizona Academy of Sciences* 6 (4): 251–53.

Felger, R. S. 1971. "Lions, Cows, and Bullets." *Defenders of Wildlife* 46 (1): 37–38.

Felger, R. S. 1971. "*Beyond the Bomb*. Review of: P. R. Ehrlich and A. H. Ehrlich, Population, Resources, Environment, W. H. Freeman, San Francisco, 1970." *Ecology* 51 (1): 192.

Felger, R. S., and C. H. Lowe. 1970. "New Combinations for Plant Taxa in Northwestern Mexico and Southwestern United States." *Journal of the Arizona Academy of Sciences* 6 (1): 82–84.

Felger, R. S., and M. B. Moser. 1970. "Seri Use of *Agave* (Century Plant)." *Kiva* 35 (4): 159–67.

Felger, R. S. 1970. "Review of J. W. Krutch, The Forgotten Peninsula: A Naturalist in Baja California, William Morrow and Co., New York, 1970." *Defenders of Wildlife* 456 (4): 428.

Felger, R. S. 1970. "Arizona and California Cacti. Reviews of: L. Benson, *The Cacti of Arizona*, University of Arizona Press, Tucson, 1969, and L. Benson, *The Native Cacti of California*, Stanford University Press, 1969." *Ecology* 51 (5): 945–46.

Moran, R., and R. S. Felger. 1968. "*Castela polyandra*, a New Species in a New Section: Union of *Holacantha* with *Castela* (Simaroubaceae)." *Transactions of the San Diego Society of Natural History* 15 (4): 31–40.

Felger, R. S., and C. H. Lowe. 1967. "Clinal Variation in the Surface-Volume Relationships of the Columnar Cactus *Lophocereus schottii* in Northwestern Mexico." *Ecology* 48 (4): 530–36.

Felger, R. S. 1965. "*Xantusia vigilis* and Its Habitat in Sonora, Mexico." *Herpetologica* 21:146–47.

# CONTRIBUTORS

**James Aronson** is president of the Ecological Health Network and an emeritus scientist with Missouri Botanical Garden. He worked for twenty-five years in the Center for Functional and Evolutionary Ecology, in Montpellier, France, and seven years at the Boyko Center for Applied Research in Beer Sheva, Israel. He has participated in and advised on ecological restoration and rehabilitation projects, especially drylands. He has co-authored or edited over three hundred articles and many influential books on ecological restoration and biology and wildlife of the Mediterranean region.

**Thomas Bowen** is professor emeritus in the Department of Anthropology at California State University, Fresno, and a research associate at the University of Arizona Southwest Center. His research on the archaeology and history of the islands in the Gulf of California is described in two books and a monograph (*Unknown Island*, 2000; *The Record of Native People on Gulf of California Islands*, 2009; and *On Desert Shores*, 2022). An avid mountaineer, in 2013 he co-discovered a population of native mice living at more than 20,000 feet in the Chilean Andes, now recognized as the highest vertebrates on earth.

As a child, **Bill Broyles** fell under the Sonoran Desert's spell. Though he majored in philosophy and English at the University of Arizona, his articles and books feature the desert and its people. His book collaborations include

*Sunshot: Peril and Wonder in the Gran Desierto*; *Last Water on the Devil's Highway*; *Among Unknown Tribes: Rediscovering the Photographs of Explorer Carl Lumholtz*; *Flying for the U.S. Border Patrol, 1940–2005*; *Remembering Ranches Around Ajo, Arizona*; and, with Richard Felger, *Dry Borders: Great Natural Reserves of the Sonoran Desert*. Broyles is affiliated with the University of Arizona's Southwest Center.

**Alberto Búrquez** is a researcher at the National University of Mexico (UNAM) since 1989 and holds a PhD from the University of Cambridge in England. He has authored numerous research articles and several books in both Spanish and English. His primary research interests encompass population ecology and genetics, mutualistic interactions among species, conservation, and the resource use and land transformation by humans in both Indigenous and non-Indigenous communities.

Botanist, naturalist, editor, and poet **Susan Davis Carnahan** lives in the grasslands of southeast Arizona. She is co-author with Richard Felger and Jesús Sánchez of *Oasis at the Desert Edge: Flora of Cañón del Nacapule, Sonora, Mexico* and *The Desert Edge: Flora of the Guaymas–Yaqui Region of Sonora, Mexico*.

**Mark A. Dimmitt** has a PhD in biology (herpetology) from the University of California at Riverside. He worked at the Arizona-Sonora Desert Museum from 1979 to 2011, first as curator of botany, and later as director of natural history. His areas of research included botany and vertebrate biology, and he is the author of more than fifty scientific and popular publications, including *A Natural History of the Sonoran Desert* (2000). He is a Fellow of the Cactus and Succulent Society of America. Mark is also an expert plant hybridizer and collects and grows a number of other weird plants, mostly succulents and epiphytes.

**Exequiel Ezcurra** has devoted his career to the study of the ecosystems of northwestern Mexico. He has published more than three hundred papers and books and directed many popular and award-winning films. He was honored with a Conservation Biology Award and a Pew Fellowship in Marine Conservation; he was scientific chair of the CITES Convention, president of Mexico's National Institute of Ecology, and eleven-year director of the University of California Institute for Mexico and the United States (UC MEXUS).

Currently, he is emeritus professor of ecology at the University of California, Riverside.

**Gil Gillenwater** is a lifelong Arizona resident. He worked in real estate and in 1987 founded Rancho Feliz, providing housing, education, food, medical services, and other resources on the border in the Agua Prieta region. Since inception, Rancho Feliz has invested over $20 million and hosted over 27,000 volunteers. Gil has visited more than seventy countries, including three expeditions into Tibet's Hidden Lands. He holds a black belt in kenpo karate and is a certified hot yoga instructor and a lifelong meditation practitioner. When not volunteering or traveling, Gil divides his time between Scottsdale and Flagstaff, Arizona, and Agua Prieta and San Carlos, Sonora, accompanied by his wife, Izuru.

**Russ Kleinman** practiced general surgery for twenty-seven years before retiring and becoming a botanist. He teaches plant taxonomy at Western New Mexico University in Silver City. Russ has studied the bryophytes of New Mexico for about the past fifteen years and has recently collaborated with Karen Blisard and Kelly Allred, resulting in the publication of the first illustrated book on the bryophytes of New Mexico.

**Angelina Martínez-Yrízar** is a researcher at the Instituto de Ecología, Universidad Nacional Autónoma de México (UNAM) in Hermosillo, Sonora. As an ecosystem ecologist, her research centers on understanding the structure and functioning of biomass, primary productivity, decomposition and nutrient cycling, and the resilience of water-limited ecosystems in the Sonoran Desert and tropical deciduous forests in Mexico.

**Gregory McNamee** is a writer, editor, and publisher who has long made his home in Tucson. He is the author of more than forty books and more than ten thousand periodical pieces.

**Cathy Moser Marlett** has contributed to studies of the Seri language and culture, drawing on her early years living in the Seri community and her ongoing research there as an adult. With a degree in art and biology, she has authored numerous publications and illustrated others. Her book *Shells on a Desert Shore* (2014) blends her multiple interests.

**Gary Paul Nabhan** worked for Richard Felger as a research assistant at the Arizona-Sonora Desert Museum from the summer of 1976 until 1980, but continued to publish with him for another forty years. Also known as Brother Coyote—an interfaith Franciscan Brother—he has three recent books out that complement Richard's work: *Agave Spirits*, coauthored with David Suro; *Against the American Grain: A Borderlands History of Resistance*; and the *Chile, Clove and Cardamom* desert cookbook with Beth Dooley. He is now spearheading the Sacred Plants Biocultural Recovery Initiative as a Takreem Foundation Laureate in the U.S.-Mexico Desert Borderlands and in the Middle East.

**Wallace J. Nichols** (1967–2024) described himself as a "marine biologist, author, water lover, turtle nerd, explorer, movement builder, embarrassing Dad and creator of useful words." His work with sea turtles included tracking one named Adelita seven thousand miles across the Pacific by satellite transmitter, and he fundraised to protect turtle nesting beaches that saved a million sea turtle hatchings. He created a movement with his best-selling book *Blue Mind*, and he gave away blue marbles representing "our beautiful, blue water planet Earth" as reminders of the therapeutic effects of spending time near, on, in, or under water. His emails closed with "I wish you water."

**William R. Norris** is professor of biology at Western New Mexico State University (since 2001), where he teaches ten-plus different classes in topics ranging from botany, fungi, algae, ecology, and ornithology to biostatistics. His research, conducted in both New Mexico and Iowa, focuses on floristics studies and vegetation monitoring of natural vegetation remnants and *Carex* taxonomy.

**Humberto Romero Morales** is the botanical expert of the Comcaac (Seri people). His work *Plant Life of a Desert Archipelago* (2012) with Richard Felger and Ben Wilder records the flora of the islands of the Gulf of California, where Humberto has lived and worked his whole life. An avid teacher, his current efforts include the creation of Jardín Botánico Xaaslca Án (The Garden of Many Columnar Cacti) outside of Punta Chueca.

**Sue Rutman** is a retired National Park Service botanist at Organ Pipe Cactus National Monument and a research associate with the University of Arizona's

Herbarium, and she writes about the hidden histories of places and people in the Southwest.

**Silke Schneider** left home in Germany, barely out of high school, in the late 1970s, with ninety horses, twenty elephants, numerous tigers and lions, dogs and other animals. How? She ran off with a traveling circus. Life has been an adventure ever since, and a lot of time a "circus" of various kinds. In 2004 she was elected a Fellow of the Royal Geographical Society in London. She now lives in her beloved New Mexico, where she moved in 2007 from Arizona with her husband, friend, and soulmate, Richard Felger.

**Jeffrey Seminoff** is leader of the Marine Turtle Ecology and Assessment Program at NOAA-Southwest Fisheries Science Center in La Jolla, California. Since 1992 Jeffrey has been involved in ecological research and conservation of sea turtles. He has a PhD from the University of Arizona (2000) and was a post-doc at the Archie Carr Center for Sea Turtle Research at the University of Florida before joining NOAA (in 2002). Seminoff is a member of the IUCN Marine Turtle and IUCN Tortoise and Freshwater Turtle Specialist Groups and is past president of the International Sea Turtle Society (2011). His research uses innovative approaches such as stable isotope analysis, biotelemetry, and aerial surveys to study sea turtles around the world. Jeff; his wife, Jennifer; daughter, Quin; and son, Graeson, enjoy sharing meals and laughter whenever the kids are in town, often while being watched by their two pugs, Arlo and Boo.

**Jane Spinti** was a counselor at the University of Wisconsin and later at New Mexico State University. She became interested in desert plants and finished a degree in horticulture while at NMSU. When she moved to Silver City, she began working in the NMSU Herbarium and hiking in a group that included Richard Felger, a great hiker and teacher.

Born and raised in the Sonoran Desert and mentored by Richard Felger from a young age, **Benjamin T. Wilder** is a biogeographer, botanist, and desert ecologist. His research looks at the origin and future of the desert, and he is co-founder and director of the Next Generation Sonoran Desert Researchers. His efforts strive to combine multiple disciplines and perspectives to better understand and protect the people and places that drive his passion. He lives with his wife, Dr. Erin Riordan, and their son, Eli, in the Tucson Mountains.

## The Southwest Center Series

JEFFREY M. BANISTER, EDITOR

Ignaz Pfefferkorn, *Sonora: A Description of the Province*

Carl Lumholtz, *New Trails in Mexico*

Buford Pickens, *The Missions of Northern Sonora: A 1935 Field Documentation*

Gary Paul Nabhan, editor, *Counting Sheep: Twenty Ways of Seeing Desert Bighorn*

Eileen Oktavec, *Answered Prayers: Miracles and Milagros Along the Border*

Curtis M. Hinsley and David R. Wilcox, editors, *Frank Hamilton Cushing and the Hemenway Southwestern Archaeological Expedition, 1886–1889*, volume 1: *The Southwest in the American Imagination: The Writings of Sylvester Baxter, 1881–1899*

Lawrence J. Taylor and Maeve Hickey, *The Road to Mexico*

Donna J. Guy and Thomas E. Sheridan, editors, *Contested Ground: Comparative Frontiers on the Northern and Southern Edges of the Spanish Empire*

Julian D. Hayden, *The Sierra Pinacate*

Paul S. Martin, David Yetman, Mark Fishbein, Phil Jenkins, Thomas R. Van Devender, and Rebecca K. Wilson, editors, *Gentry's Rio Mayo Plants: The Tropical Deciduous Forest and Environs of Northwest Mexico*

WJ McGee, *Trails to Tiburón: The 1894 and 1895 Field Diaries of WJ McGee*, transcribed by Hazel McFeely Fontana, annotated and with an introduction by Bernard L. Fontana

Richard Stephen Felger, *Flora of the Gran Desierto and Río Colorado of Northwestern Mexico*

Donald Bahr, editor, *O'odham Creation and Related Events: As Told to Ruth Benedict in 1927 in Prose, Oratory, and Song by the Pimas William Blackwater, Thomas Vanyiko, Clara Ahiel, William Stevens, Oliver Wellington, and Kisto*

Dan L. Fischer, *Early Southwest Ornithologists, 1528–1900*

Thomas Bowen, editor, *Backcountry Pilot: Flying Adventures with Ike Russell*

Federico José María Ronstadt, *Borderman: Memoirs of Federico José María Ronstadt, edited by Edward F. Ronstadt*

Curtis M. Hinsley and David R. Wilcox, editors, *Frank Hamilton Cushing and the Hemenway Southwestern Archaeological Expedition, 1886–1889*, volume 2: *The Lost Itinerary of Frank Hamilton Cushing*

Neil Goodwin, *Like a Brother: Grenville Goodwin's Apache Years, 1928–1939*

Katherine G. Morrissey and Kirsten Jensen, editors, *Picturing Arizona: The Photographic Record of the 1930s*

Bill Broyles and Michael Berman, *Sunshot: Peril and Wonder in the Gran Desierto*

David W. Lazaroff, Philip C. Rosen, and Charles H. Lowe Jr., *Amphibians, Reptiles, and Their Habitats at Sabino Canyon*

David Yetman, *The Organ Pipe Cactus*

Gloria Fraser Giffords, *Sanctuaries of Earth, Stone, and Light: The Churches of Northern New Spain, 1530–1821*

David Yetman, *The Great Cacti: Ethnobotany and Biogeography*

John Messina, *Álamos, Sonora: Architecture and Urbanism in the Dry Tropics*

Laura L. Cummings, *Pachucas and Pachucos in Tucson: Situated Border Lives*

Bernard L. Fontana and Edward McCain, *A Gift of Angels: The Art of Mission San Xavier del Bac*

David A. Yetman, *The Ópatas: In Search of a Sonoran People*

Julian D. Hayden, *Field Man: The Life of a Desert Archaeologist*, edited by Bill Broyles and Diane Boyer

Bill Broyles, Gayle Harrison Hartmann, Thomas E. Sheridan, Gary Paul Nabhan, and Mary Charlotte Thurtle, *Last Water on the Devil's Highway: A Cultural and Natural History of Tinajas Altas*

Thomas E. Sheridan, *Arizona: A History, Revised Edition*

Richard S. Felger and Benjamin T. Wilder, *Plant Life of a Desert Archipelago: Flora of the Sonoran Islands in the Gulf of California*

David Burkhalter, *Baja California Missions: In the Footsteps of the Padres*

Guillermo Núñez Noriega, *Just Between Us: An Ethnography of Male Identity and Intimacy in Rural Communities of Northern Mexico*

Cathy Moser Marlett, *Shells on a Desert Shore: Mollusks in the Seri World*

Rebecca A. Carte, *Capturing the Landscapes of New Spain: Baltasar Obregón and the 1564 Ibarra Expedition*

Gary Paul Nabhan, editor, *Ethnobiology for the Future: Linking Cultural and Ecological Diversity*

James S. Griffith, *Saints, Statues, and Stories: A Folklorist Looks at the Religious Art of Sonora*

David Yetman, Alberto Búrquez, Kevin Hultine, and Michael Sanderson, with Frank S. Crosswhite, *The Saguaro Cactus: A Natural History*

Carolyn Niethammer, *A Desert Feast: Celebrating Tucson's Culinary Heritage*

Gary Paul Nabhan, editor, *The Nature of Desert Nature*

David Yetman, *Natural Landmarks of Arizona*

Michael Chiago Sr. and Amadeo M. Rea, *Michael Chiago: O'odham Lifeways Through Art*

David Wentworth Lazaroff, *Picturing Sabino: A Photographic History of a Southwestern Canyon*

David Yetman and Alberto Búrquez, *Mexico's Valleys of Cuicatlán and Tehuacán: From Deserts to Clouds*

Curtis M. Hinsley and David R. Wilcox, editors, *Frank Hamilton Cushing and the Hemenway Southwestern Archaeological Expedition, 1886–1889*, volume 3: *On a Trail of Southwest Discovery: The Expedition Diaries of Frederick W. Hodge and Margaret W. Magill, 1886–1888*

David Burckhalter, *Baskets from the Seri Coast: Comcaac Weavers and Their Craft*

Benjamin T. Wilder, Bill Broyles, and Thomas Bowen, editors, *The Botanical Odyssey of Richard S. Felger*

Printed and bound by CPI Group (UK) Ltd, Croydon, CR0 4YY

13/08/2026

14941848-0001